Urban Ecosystem Justice

Merging together the fields of urban ecology, environmental justice, and urban environmental education, *Urban Ecosystem Justice* promotes building fair, accessible, and mutually beneficial relationships between citizens and the soils, water, atmospheres, and biodiversity in their cities.

This book provides a framework for recentering issues of justice and fairness in sustainability discourse while challenging the profound ecological alienation experienced by urban residents. While the urban sustainability movement has had many successes in the past few decades, there remain areas for it to grow. For one, the benefits of sustainability have disproportionately benefited wealthier city residents, with concerns over equity, justice, and social sustainability frequently taking a back seat to economic and environmental considerations. Additionally, many city dwellers remain estranged from and unfamiliar with ecological processes, with urban environments often thought of as existing outside of nature or as hopelessly degraded. Through a citizen-centered lens, the book offers a guide to reconciling these issues by demonstrating how questions of equity, access, and justice apply to the biophysical dimensions of the urban ecosystem: soil, water, air, waste, and biodiversity. Drawing heavily from the fields of urban ecology, environmental justice, and ecological design, this book lays out a science of cities for people: a pedagogical platform that can be used to promote ecological literacy in underrepresented urban communities through affordable and decentralized means.

This book provides both a theoretical and practical field guide to students and researchers of urban sustainability, city planners, architects, policymakers, and activists wishing to develop reciprocal relationships with urban ecologies.

Scott Kellogg has been Educational Director of the Radix Ecological Sustainability Center, an urban environmental education and just sustainabilities advocacy nonprofit based in Albany, New York, since 2009. Prior to Radix, Scott was a cofounder and collective member of the Rhizome Collective in Austin, Texas, an urban sustainability and activist center which functioned from 2000 to 2009. Scott has a PhD in Science and Technology Studies from Rennselaer Polytechnic Institute, a master's degree in Environmental Science and Policy from Johns Hopkins, and a bachelor's degree in Humanities from the New College of California. He teaches in a number of graduate and undergraduate programs in universities throughout the Northeast, including Bard College's Masters in Environmental Education program. Scott is the coauthor of *Toolbox for Sustainable City Living: A Do-It-Ourselves Guide* (2008).

Routledge Equity, Justice and the Sustainable City

Series editors: Julian Agyeman and Stephen Zavestoski

This series positions equity and justice as central elements of the transition toward sustainable cities. The series introduces critical perspectives and new approaches to the practice and theory of urban planning and policy that ask how the world's cities can become 'greener' while becoming more fair, equitable and just.

The *Routledge Equity Justice and the Sustainable City* series addresses sustainable city trends in the global North and South and investigates them for their potential to ensure a transition to urban sustainability that is equitable and just for all. These trends include municipal climate action plans; resource scarcity as tipping points into a vortex of urban dysfunction; inclusive urbanization; "complete streets" as a tool for realizing more "livable cities"; the use of information and analytics toward the creation of "smart cities".

The series welcomes submissions for high-level cutting edge research books that push thinking about sustainability, cities, justice and equity in new directions by challenging current conceptualizations and developing new ones. The series offers theoretical, methodological, and empirical advances that can be used by professionals and as supplementary reading in courses in urban geography, urban sociology, urban policy, environment and sustainability, development studies, planning, and a wide range of academic disciplines.

Understanding Urban Cycling
Exploring the Relationship Between Mobility, Sustainability and Capital
Justin Spinney

Understanding Just Sustainabilities from Within
A Case Study of a Shared-Use Commercial Kitchen in Connecticut
Phoebe Godfrey

Urban Ecosystem Justice
Strategies for Equitable Sustainability and Ecological Literacy in the City
Scott Kellogg

For more information about this series, please visit www.routledge.com/Routledge-Equity-Justice-and-the-Sustainable-City-series/book-series/EJSC

Urban Ecosystem Justice

Strategies for Equitable Sustainability and Ecological Literacy in the City

Scott Kellogg

Routledge
Taylor & Francis Group
LONDON AND NEW YORK

First published 2022
by Routledge
2 Park Square, Milton Park, Abingdon, Oxon OX14 4RN

and by Routledge
605 Third Avenue, New York, NY 10158

Routledge is an imprint of the Taylor & Francis Group, an informa business

British Library Cataloguing-in-Publication Data
A catalogue record for this book is available from the British Library

Library of Congress Cataloging-in-Publication Data
Names: Kellogg, Scott T., author.
Title: Urban ecosystem justice : strategies for equitable sustainability and ecological literacy in the city / Scott Kellogg.
Description: Milton Park, Abingdon, Oxon ; New York, NY : Routledge, 2022. | Includes bibliographical references and index.
Identifiers: LCCN 2021014786 (print) | LCCN 2021014787 (ebook)
Subjects: LCSH: Urban ecology (Sociology) | Sustainable development. | Environmental justice.
Classification: LCC HT241 .K46 2022 (print) | LCC HT241 (ebook) | DDC 307.76—dc23
LC record available at https://lccn.loc.gov/2021014786
LC ebook record available at https://lccn.loc.gov/2021014787

ISBN: 978-0-367-85870-4 (hbk)
ISBN: 978-1-032-08047-5 (pbk)
ISBN: 978-0-367-85871-1 (ebk)

DOI: 10.4324/9780367858711

Typeset in Goudy
by Apex CoVantage, LLC

Contents

Figures and tables

Figures

Tables

Author's prologue: sustainability – the new opioid of the masses

Since the beginning of the new millennium I have built a professional career in the field of urban sustainability. While consisting in part of consultation, writing, lecturing, and policy work, the bulk of it has been in the nonprofit sector, designing, managing, and operating urban sustainability demonstration and education sites. These have been the Rhizome Collective in Austin, Texas, which functioned from 2000 to 2009 and the Radix Ecological Sustainability Center in Albany, New York, which began in 2009 and continues to operate to the present day. During this time, the shape, purpose, and central players of urban sustainability have changed enormously, evolving from a radical fringe movement rife with transformative potential to a thoroughly mainstream, co-opted, misapplied, and potentially dangerous concept. This book is the outgrowth of these years of experience and research and an effort to grapple with the current state and future direction of the urban sustainability concept.

My personal path from early environmental consciousness to the formation of the Rhizome Collective has been one of change and transformation, as well. While I was in high school, I considered myself to be an "environmentalist", concerned with the most prominent environmental issues of the time including rainforest destruction, ozone depletion, acid rain, toxic waste, the greenhouse effect, and so on. These "post-Earth Day" causes popular in the late 1980s and early 1990s took place before sustainability discourse became widespread and reinforced a modernistic notion that humans were an inherently destructive species separate from nature and destined to ruin the planet. The message internalized was that the best an environmentalist could hope to do would to be less of a human: conserve resources, be a vegetarian, recycle, consume less, avoid procreation, feel guilty about one's own existence, etc. Taken to its logical conclusion, this mindset would argue that it would be best for humanity to commit speciecide in order to save the earth. Seldom, if ever, discussed were ways in which it would be possible for humans to live in a relative, if only discordant, harmony with the global ecosystem.[1] We remained ignorant of the many examples of human societies doing so outside of the trajectory of Indo-European agriculturalists-turned-dominant Western culture with which we were familiar (Balée, 2002).

These attitudes were strongly reinforced for me at a Model UN Conference I attended in eleventh grade at the United Nations where a speaker discussed rainforest conservation in the Amazon Basin and the need to create protected preserves free of human activity (including indigenous peoples). Years later (in my PhD program) I read a piece by Donna Haraway where she deftly dissects this standpoint, describing it as an "Eden under glass" (Jamison, 1992). At some point in my early undergraduate years as my personal political beliefs became more radical, I became disenchanted with the environmental movement, seeing it as a weak liberal cause that did little to significantly challenge the capitalistic forces responsible for ecological destruction and social injustice. I became more involved with decentralized nonviolent anarchist "social" campaigns such as Food Not Bombs, Homes Not Jails, and political prisoner campaigns. Involvement with these radical movements helped me to develop a solid critique of capitalist society, one that has remained central, albeit less pronounced, in my work to this day.

I remained involved in radical causes throughout my post-undergraduate years, culminating in the then-nascent global justice movement's massive demonstrations against entities such as the IMF, WTO, the World Bank, and the US Republican party. While initially exhilarating, these rallies ultimately led to me suffering from a degree of activist burnout – multiple arrests as well as the effects of low-intensity psychological warfare methods employed by police necessitated that I changed my personal tactics for achieving social transformation. The alter-globalization movement had a clear stance and well-developed critique against global capitalism but offered little in the way of solutions. I began to ponder the meaning of the movement's rallying cry "another world is possible": How do people eat in this other world? How do they procure water, energy, and deal with their wastes? How are the roles and responsibilities typically carried out by governmental agencies such as education, health care, transportation, and decision-making fulfilled by community organizations? Is it possible to plant the seed of this alternative world while simultaneously taking a hatchet to the dominant systems which we opposed?

Pondering these questions sent me on a quest to find their answers that led me to the permaculture movement. Permaculture (a portmanteau of "permanent" and "agriculture" and "culture") is a design philosophy that strives to create regenerative human–ecological relationships in areas including (but not limited to) food, water, waste, building, energy, and social systems. While drawing primarily from the ecological sciences, permaculture also borrows concepts from traditional ecological knowledge, agronomy, landscape design, architecture, and a variety of other disciplines, giving students a practical skill set of tools and technologies for building sustainable infrastructures (Lockyer, 2013). While not explicitly politically aligned in one sense or another, permaculture's open source and decentralized structure made it highly compatible with the radical anarchistic philosophies I held at the time. In this sense, permaculture provided the philosophical and pragmatic counterbalance to my activism-informed social criticism that I had been seeking – for every "no" I previously had, I now had a "yes". Being able to point to positive, functioning alternatives lent considerable strength and validity to my critique.

While having enormous potential as both a design philosophy and a worldview, the permaculture movement was not without its shortcomings. Foremost among them is the fact that permaculture was created primarily for extensive rural and land-based applications. While employed by many working farmers in the Global South, based on my observations most of permaculture's adherents in the Global North were white, middle class, and landowning. While there was some discussion of urban applications, most of the material in a standard permaculture curriculum had little relevance to city dwellers and even less to low-income urban renters. Furthermore, I saw permaculture's political ambiguity as a potential weakness of the movement (while mostly dominated by leftists, there was a troubling right-leaning libertarian element at play as well).

In response to these critiques, I sought to synthesize an urban-based permaculture with anarchist politics, developing a concept I named "radical sustainability". This was also a reaction to the growing neoliberalization of sustainability (most notably in the form of "sustainable development") and to corporate greenwashing in general (Swyngedouw, 2010). With radical sustainability in mind, myself and a handful of other like-minded activists cofounded the Rhizome Collective in Austin, Texas. Rhizome consisted of a nearly 10,000 square foot warehouse that functioned both as an urban sustainability demonstration site and as a center for community organizing, providing space to various activist organizations including Food Not Bombs, Bikes Across Borders, the Independent Media Center, and Inside Books (a books-for-prisoners operation). Interspersed among these organizations' workspaces were gardens, constructed wetlands, wind turbines, chickens, parabolic solar cookers, rain barrels, aquaponics fish farms, and composting toilets. Intended to be simple, affordable, and made of repurposed components, the placement and respective ebbs and flows of these systems were strategically aligned, guided by permaculture design principles (Holmgren, 2002).

By housing these disparate components in common space, we hoped to create a vital cross-pollination of people, ideas, and interests, tangibly demonstrating the necessary interconnectedness of social and ecological justice issues (and therein challenging a persistent meme in many activist circles that maintains these as separate). For us, "rhizome" was both a literal botanical term and a metaphor. Inspired by the Deleuzian/Guattarian concept of rhizomatic organization (Deleuze, 2004), we saw ourselves as being one autonomous chapter that was linked to others at the root level through an interconnected network. Just like the running bamboo that was spread around the space's periphery, our movement was also hidden underground, tenacious, and difficult to uproot.

The Rhizome Collective lasted from 2000 to 2009, in that time evolving from a burned-out dilapidated shell of a warehouse inhabited by circus punks to a reasonably well-functioning consensus-run collective, operating a thriving community space and receiving federal grant funding. Rhizome's peak likely occurred in 2004 with the acquisition of a ten-acre brownfield parcel and subsequent receipt of an EPA Brownfields Cleanup Award. The property, located on the edge of Austin city limits, had been previously used for several decades as an illegal dumping site and was covered in a literal mountain of debris. Aware of how much it would cost to clean the site, its owner was eager to donate it to a nonprofit and at least receive a tax write-off. Upon learning of the property, we simultaneously applied for and received a $200,000 EPA grant to clean the site and transform it into an urban sustainability education park. Over the next two years we set about the cleanup, paying ourselves to carefully remove and separate out every piece of garbage, sorting them according to their reuse or recyclability potential. State and federal authorities were stunned when we finished the project on schedule and received a certificate of completion, amazed that we were able to complete the cleanup with such limited funding and scant resources. It was likely our nimbleness, resourcefulness, motivation, and lack of extensive overhead costs that allowed us to succeed where larger organizations failed.[2]

Perhaps, the most interesting aspect of the entire project to me was our attendance of two annual national EPA-sponsored brownfield conferences in Denver and Boston. Expecting to find other small nonprofits such as ours working on similar endeavors, I instead found representatives of an apparently large and growing industry related to the acquisition, cleanup, and subsequent redevelopment of blighted urban properties. Comprising a coalition of government officials, massive engineering firms, and real estate interests, it became obvious to me that government brownfield programs primarily served to redirect federal dollars into these private corporations, who no doubt in turn created a powerful lobby. A handful of small nonprofits such as ours were given funding to create the appearance that issues of social sustainability were taken into consideration. At these conferences there was virtually no discussion about the social justice components or potential negative repercussions of brownfields development, such as creating local employment or the risk of gentrification. That was perhaps my first exposure to a style of depoliticized technocratic sustainability governance that I've since encountered many times and that in many ways has come to dominate much of sustainability discourse.

In the time between the creation of Rhizome and today, the character and extent of urban sustainability, along with its associated discourses, have changed considerably. What was a radical idea in 2000 has since become commonplace, with the idea of "sustainable cities" now having largely been embraced by urban planners, corporations, and the mainstream media (Swyngedouw, 2010). When Rhizome began, it seemed to me (perhaps naively) that merely planting a garden in the city was a radical act. By 2008, it was obvious that it was no longer the case. Sensing that urban sustainability was in danger of being co-opted, Rhizome cofounder Stacy Pettigrew and I wrote the book, *Toolbox for Sustainable City Living: A Do-It-Ourselves Guide* (Kellogg, 2008). The book was an attempt not only to document the previous years' work at Rhizome but also to securely situate urban sustainability (or at least aspects of it) within a radical political platform. It was here that we detailed the notion of "radical sustainability" and made explicit references to its use in "dual-power anarchist strategies". Furthermore, the book extensively critiqued "green capitalism" and warned of the potential for sustainability endeavors to result in the gentrification of low-income communities. Being one of the first books of its kind on a topic of growing popularity, "Toolbox" was well received and sold vigorously. Within a year of its publication, however, a flurry of other similarly themed "urban homesteading" books were released.

In general, they were apolitical, slickly produced, and no doubt appealed to a wider commercial audience. Urban sustainability had gone mainstream, and "Toolbox" was quickly relegated to radical fringes. Immediately, I witnessed a backlash from "post-politically" minded sustainability advocates. At one book publicity event in Phoenix, Arizona, I was instructed by the organizer to avoid bringing up any politically controversial topics and to only focus on "gee-whiz" type positive solutions (I proceeded to anyway, much to their chagrin). I saw a dangerous potential for sustainability to become an opioid of the masses. Promising simple solutions – pretty gardens, solar panels, bicycle transportation, and rain barrels – people were distracted, feeling as if they were making a difference all the while the capitalist machine churned onwards. Permaculture, if not continually engaged with critical issues of equity, justice, and access, risked being reduced to a chic gardening fad of the middle class.

Rhizome ultimately fell victim to the pressures of urban growth that we were unintentionally complicit in fueling. In 2009, an internal policy change in Austin's City Code enforcement department triggered building inspectors to target spaces such as ours that informally hosted arts, music, or community activism. Spaces such as these were apparently seen as undesirable to the new incoming occupants of the high-end residential complexes now being built across the city's historically Latino East Side. Even the industrially zoned neighborhoods where Rhizome was located were subject to crackdowns. This process of displacement and gentrification had been underway for some time, however, and was to at least some degree exacerbated by the presence of "bohemian" operations such as Rhizome. Despite being outspoken against the rapid increase in property values and allying with anti-gentrification causes, we ultimately succumbed to the enormous economic pressure put on our community, a consequence of Austin's incredible growth rate over the 2000–2010 decade. In March 2009, we were subject to a building inspection that resulted in extensive code violations that would have been impossibly expensive to remedy. Somewhat ironically, many of the innovative sustainability features at Rhizome that had won us widespread acclaim from the city were considered among the most egregious violations. Facing extensive penalties, I was left with no choice but to sell the building that housed the Rhizome Collective.

On the heels of the destruction of Rhizome, I moved with my wife and two daughters to Albany, New York, to begin again. Albany and Austin are two very different cities. Apart from both being capital cities that begin with the letter "A", they have very little in common in terms of climate, history, culture, politics, population, economic processes, social dynamics, or otherwise. While Austin is a young, rapidly growing (sprawling) typical sunbelt city that celebrates its "weirdness", Albany is an old shrinking post-industrial "rust belt" town with entrenched politicians, cold winters, and in further contrast to Austin, celebrates being "boring". While moving to Albany may have seemed like an unlikely choice, there were reasons why it was chosen over other places – as both my wife and I had family in the region, we had been spending summers in Albany for several years already. In "shrinking cities" (a phenomenon I discuss later in this book), there are opportunities to address both affordable housing and food security issues simultaneously. Unlike in fast-growing cities where every square foot of space is contested, the abundance of housing stock and vacant property in shrinking cities makes it possible to have both affordable homes and urban agriculture. Furthermore, the challenge of repurposing old industrial infrastructure for sustainable uses is a monumental and fascinating challenge that must be carried out throughout the world in cities like Albany.

With these considerations in mind, we conceived the idea of the Radix Ecological Sustainability Center. Radix (the Latin word for "root") clearly had many parallels with Rhizome in both its name and purpose; however, there would be key differences. With Rhizome, I commonly felt that I was spread too thin by managing its many moving parts and was never successful in developing serious environmental educational programming for youth (the freaky post-apocalyptic aesthetic of Rhizome certainly did not help parents or teachers to feel comfortable in bringing their children there, either). With Radix, I wanted to focus on being an environmental education center that existed primarily to serve inner-city children and youths, building meaningful and long-lasting relationships with area public schools. Building this level of community trust would not be an easy process, but I felt compelled to begin.

From 2009 to the present day, I have been working as the educational director of the Radix Center. I have been successful at creating a functioning urban environmental education center consisting of a one-acre farm containing a solar greenhouse, microlivestock, aquaponics, bees, gardens, perennial food forest, composting operation, and more. We receive numerous visitations each year from groups ranging from elementary school age children to university students. With each group, I try to increase their ecological literacy and demonstrate how green and agricultural spaces function when integrated into a high-density urban mosaic.

Since the publication of "Toolbox" and the beginning of Radix, I have been trying to push the boundaries of what is commonly regarded as pertinent to urban sustainability: both in being more explicit regarding issues of equity and justice and in "going beyond the backyard". A serious inherent limitation in the idea of "urban homesteading" is that it makes sustainability primarily about the individual. The allusion to "homesteading" alone conjures sentimental images of the rugged frontiersman, self-reliant and self-sufficient. In many ways, self-sufficiency is antithetical with urban living: urban residents, historically, are specialists as opposed to generalists and can rely on the skills and expertise of their neighbors through barter and exchange. In many ways, the greatest perk and asset of living in a city is the freedom to not have to be a jack-of-all-trades but rather to become particularly proficient in one area. In truth, I am only so critical as I am fully in support of urban residents in being more self-reliant, however, I would encourage them to recognize the importance of social interconnectivity and the potential of collective information exchange within an urban system.

I was additionally interested in looking at urban ecological systems beyond the level of the individual homestead, backyard, lot, garden, etc. What was the synergistic effect of all these spaces interacting with each other? What could we say about creating mutually reciprocal relationships with urban waterways, soils, and atmospheres that cut across and beyond the level of the individual sustainability site? What were our interactions like with the non-human living systems within a city that we were not directly cultivating? How much of a barrier do toxic industrial legacies create to furthering these relationships? Could we look at urban permaculture on a whole systems scale "beyond the backyard"?[3]

Asking these questions led me quickly to the discipline of urban ecology, a field that was and is almost entirely studied from a top-down perspective and is made up of primarily by experts in the fields of urban planning and design. I began to wonder if the insights of this discipline had potential to be used by citizen sustainability activists; could obscure ideas such as urban metabolisms (Girardet, 1992), patch dynamics (Alberti, 2005), airsheds (Kirstine), anthrosols (Smith, 1980), novel ecosystems (Robbins, 2014), synanthropic species (Rodewald, 2008), and reconciliation ecology (Rosenzweig, 2003) be appropriated to be part of their toolkit? Furthermore, I wanted to apply questions of social justice, access, and equity to the urban ecosystem concepts that are seldom more than superficially considered in professional circles. Who has access to the rivers that flow through cities and what barriers stand between them? Which populations are most disproportionately exposed to soilborne toxins, and can they proactively work to reduce them? How do we, as citizens, think about our urban atmospheres? Similar questions have been raised numerous times by the environmental justice movement but most commonly from a "reactive" standpoint. Could we have respectful, loving, nurturing (if not imperfect) relationships with the various dimensions of the urban ecosystem? If possible, could ecosystem justice's critical analysis be combined with the ethos of urban homesteading to create mutually reciprocal symbioses between the human residents of a city and the rivers, skies, soils, birds, mammals, bugs, plants, and microbes cohabitating and cocreating them with us?

It is with these fundamental questions that I began my PhD program, the research for my dissertation, and now this book, coining the term "urban ecosystem justice" for use as an overarching framework. The privilege of being a graduate student had given me the space and access to extensive and disparate bodies of literature related to these themes. Much of my work and research has been an attempt at synthesis, combining the critique of one discipline with the insights of another. Many researchers approach the topic of urban ecosystem justice peripherally yet are not explicitly in dialogue with one another. This is most starkly noticeable in the separation between the social and natural sciences, a long-standing division that both confounds and frustrates me. In addition to the aforementioned areas of urban ecology, urban design, and environmental justice, I have also drawn significantly from the fields of political ecology, geography, critical urban theory, historical ecology, science and technology studies, and socio-ecological systems theory. Each discipline offers unique insights that help to create a fuller, more complex picture of the topic.

One important realization I've come to in my research, however, is that it's not necessary, realistic, or appropriate for me to try to develop any sort of "unifying theory of cities". Cities are inherently complex, dynamic, and constantly adapting environments made up of a plurality of moving parts and stories.

To quote Timothy Morton, "Places contain multitudes", the whole of a city is less than the sum of its parts (Morton, 2016). In this sense, Morton implies that cities are almost *anti*-emergent phenomena! From a distance, they may appear to be a monolithic whole, but once you zoom in, you realize that there are infinitely more dimensions to them at the micro level – larger on the inside than on the outside. I feel liberated in not having to understand, describe, and reduce all their processes into a single set of ideas or equations. In this sense, approaches to just urban transitions require a necessary diversity and plurality of approaches – decentralized and horizontally distributed. Sustainabilities are situated and contextualized, sharing common aspects yet uniquely tailored to the micro-political conditions of the point of their implementation (Sze, 2018).

It is with this history and these experiences that I present this book. It is my wish that it may serve as a toolbox of concepts and practical skills useful to the activist-scholar working for just sustainabilities and ecological equities and that it may further the goal of transferring knowledge from the confines of the ivory tower to the intellectual commons.

Notes

1 Botkin, Daniel B. *The moon in the nautilus shell: Discordant harmonies reconsidered*. Oxford, UK: Oxford University Press, 2012.
2 See Proctor, Rachel. *Showing rhizome the money*, May, Friday, December 17, 2004. www.austinchronicle.com/news/2004-12-17/243200/

Works cited

Alberti, M. (2005). The effects of urban patterns on ecosystem function. *International Regional Science Review*, 28(2), 168–192.

Balée, W. a. (2002). *Advances in historical ecology*. New York: Columbia University Press.

Deleuze, G. (2004). *Anti-Oedipus*. London: A&C Black.

Girardet, H. (1992). *The Gaia Atlas of cities: New directions for urban living*. New York: UN Habitat.

Holmgren, D. (2002). *Principles & pathways beyond sustainability*. Hepburn: Holmgren Design Services.

Jamison, P. K. (1992). No Eden under glass: A discussion with Donna Haraway. *Feminist Teacher*, 10–15.

Kellogg, S. a. (2008). *Toolbox for sustainable city living: A do-it-ourselves guide*. Boston, MA: South End Press.

Lockyer, J. a. (2013). *Environmental anthropology engaging ecotopia: Bioregionalism, permaculture, and ecovillages* (Vol. 17). New York: Berghahn Books.

Morton, T. (2016). *Dark ecology: For a logic of future coexistence*. New York: Columbia University Press.

Robbins, P. (2014). No going back: The political ethics of ecological novelty. In *Traditional wisdom and modern knowledge for the earth's future* (Vol. 1, pp. 103–118). Japan: Springer.

Rodewald, A. D. (2008). Consumer resource matching in urbanizing landscapes: Are Synanthropic species over-matching. *Ecology*, 89(2), 515–521.

Rosenzweig, M. L. (2003). Reconciliation ecology and the future of species diversity. *Oryx, 37*(2), 194–205.

Smith, N. J. (1980). Anthosols and human carrying capacity in Amazonia. *Annals of the Association of American Geographers, 70*(4), 553–566.

Swyngedouw, E. (2010). *Impossible sustainability and the post-political condition: Making strategies in spatial planning*. Dordrecht, the Netherlands: Springer.

Sze, J. e. (2018). *Sustainability: Approaches to environmental justice and social power*. New York: New York University Press.

Introduction

As the majority of humans on the earth now live in urban environments, it could be argued that the battle for a sustainable world could be fought and won primarily within cities. Cultural leverage points, economic engines, centers of innovation, and home to millions, how these complex adaptive socio-ecological urban entities unfold will have profound impacts on the rest of the planet. Will cities become centers of progressive ideals of justice and equity where residents coexist in relative harmony with the non-human world, or will they become monstrous toxin-producing resource vacuums with entrenched and chasmic disparities between the haves and have-nots? These are critical questions, and the field of urban sustainability has been developed to inform the responses. This discipline has undergone both considerable growth and change over the past few decades. Once regarded as a radical fringe topic (albeit with revolutionary potential), the idea of sustainable cities has now become quite mainstream, with numerous books and journals dedicated to its study. Inasmuch as the concept has been increasingly co-opted by corporate and neoliberal interests, it is often reduced to a marketing buzzword. More insidiously, though, sustainability discourse is now used to create a greenish veneer to an emerging form of enviro-technocratic governance that provides premium environmental services to wealthy residents while relegating the poor to ecological sacrifice zones on the city's toxic periphery (Swyngedouw, 2009). This divide has the ill effect of instilling a sense of separateness from and fear of the non-human among the urban populace, further widening the false duality of nature and society. With pressures from persistent inequality, climate change, energy depletion, and biosphere degradation continuing to mount (Steffen, 2015), the imperative to initiate urgent just urban transformations grows greater.

It is here that urban sustainability is at a crossroads – can it be reclaimed by citizen-centered and grassroots organizations for the purpose of building genuinely just and healthy communities, or has urban sustainability become nothing more than another "opiate of the masses" to keep us distracted while neoliberal business-as-usual carries on? To inform this discussion, it is necessary to analyze what I believe to be two of the primary challenges that the urban sustainability movement now faces. These are what I refer to respectively as "social-sustainability exclusion" and "ecological alienation".

DOI: 10.4324/9780367858711-1

Social sustainability exclusion

Sustainability discourse over recent years has placed a disproportionate emphasis on the "environmental" and "economic" aspects of sustainability, largely ignoring or underemphasizing sustainability's social dimension (Agyeman, 2002). This trend has produced a form of techno-managerial sustainability or "post-politics" (Swyngedouw, 2010) that is attractive to business owners, policymakers, and the ruling classes as it promotes a "green" agenda that is at once friendly to capital and conducive to crafting the illusion of community consensus. By relegating the social component to the sidelines, inconvenient questions regarding equity, access, fairness, race, and class are glossed over, and fundamental structural socioeconomic inequalities are never addressed. As such, the status quo remains unchallenged and environmental initiatives privilege only affluent communities. Little or no attempt is made to ensure that there is equitable distribution of environmental harms and goods, and in cases when environmental amenities are provided to low-income communities, it often results in the unintentional (or intentional?) consequence of their displacement/cultural alienation (i.e., gentrification) (Gould, 2016). The most extreme form in which this manifests is the phenomenon of "urban ecological securitization" (Hodson, 2009), where premium environmental services are provided to the wealthy and the poor are displaced to the urban edges where they are subject to the brunt of ecological risk, exposure, and vulnerability. More recently, extreme visions for such techno-ecological authoritarian regimes have manifested under the name of "ecomodernism" (Asafu-Adjaye, 2015) that would propose that humans live entirely segregated from nature in self-contained urban spheres.

Ecological alienation

Ecological alienation or ecological rift (McClintock, 2010) is a present-day manifestation of the nature–society dualism professed through modernist ideals and philosophy (Descola, 1996). Through it, urban residents are profoundly separated from and ignorant of the natural processes and systems that make life on the earth possible (i.e., food production, water, composting/decomposition, energy, atmospheric/climatic processes, non-human life). The separation of town and country has relegated these processes and systems to the urban periphery or hinterlands, making them invisible and inaccessible to urban residents (Elmqvist, 2013). In instances where "remnant" ecologies remain in cities, they are commonly made inaccessible through enclosure, poisoned, degraded, or otherwise devalued. It has only been in the past few decades that the discipline of urban ecology itself has embraced the role that human and social processes play in urban ecosystem functioning, prompting a shift in emphasis from the "ecology in" to the "ecology of" cities (Pickett, 2016). Where environmental education does exist in cities, it teaches about the environment and nature as external to the city, with urban environments being considered unworthy of study (Thomashow, 2015). Likewise, definitions of "the environment" seldom are extended to include social and human processes. The combined influence of these conditions produces in both children and adults what is referred to as "ecophobia", or fear of ecological systems and processes (Sobel, 1996). When a citizenry has no sense of interrelation, love, concern, or responsibility for ecological systems, they cannot be expected to act in their defense.

Synthesis and solution: urban ecosystem justice

Urban ecosystem justice is an attempt to reconcile these two problems, and inasmuch as it hopes to redirect the energy and intent of the urban sustainability movement toward more progressive and regenerative ends. In framing cities as complex, adaptive socio-ecological systems that are the coevolutionary products of both human and non-human processes (Alberti, 2008), urban ecosystem justice looks at how questions of equity, access, fairness, race, and class apply to the biophysical dimensions of urban ecosystems (soil, water, waste, air, biodiversity). By doing so, it makes explicit and moves social sustainability to the forefront of sustainability discourse, while simultaneously challenging ecological alienation by making the urban ecosystem a legitimate, and relevant topic of study. Very importantly, urban ecosystem justice situates itself within a citizen-centered, grassroots context (Smith, 2016). In this regard, it serves as a pedagogical tool kit focused on exploring and creating mutually reciprocal symbioses between ordinary citizens and urban ecologies from the ground up, an angle typically not explored from top-down planning and policy perspectives. By applying a DIY ethic to the "ecology of cities" paradigm developed in the discipline of urban ecology (Pickett, 2016), humans are central and integral to urban environmental processes. In this regard, urban ecosystem justice can be thought of a "science of cities" for the people (McPhearson, 2016).

This book is divided into three main parts. The first part consists of a thorough analysis of the questions of ecological alienation and equity exclusion, largely based around an examination of related concepts and their reconciliatory potential. Collectively, these concepts form a "chaotic bricolage" that comprise the urban ecosystem justice mosaic. This first part is heavily theoretical and functions to unpack key concepts and analyze how they fit within and contribute to a larger urban ecosystem justice framework. The second part is an in-depth exploration of urban ecosystem justice itself in both its theoretical and practical components, divided into sections pertaining to the various biophysical "spheres" of the urban ecosystem: soil, water, waste (compost), atmospheres, and biodiversity. The second part is less theoretical than the first and delves into more applied, grounded applications and discussions of the nuts and bolts of urban ecosystem justice. The third part is a study in applied urban ecosystem justice told in the story of the Radix Ecological Sustainability Center and its educational philosophy. I hope the reader will be able to draw equally upon all parts of the text.

This book, particularly in Part II, places a strong emphasis on the practical application of this knowledge, containing a number of "how-to" sections that provide a step-by-step instruction on how to construct and implement the learning tools and systems described. While this format may be atypical for an academic work, I believe it provides a critical bridge between the realms of theory and practice and will hopefully be of great use to the scholar-activist engaged in the fields of urban environmental education, ecological design, environmental justice advocacy, and all manner of "situated solidarities" (Routledge, 2015). Like the false dichotomy of nature and society, the mythological divide between activist and scholar is similarly damaging – too often academics will theorize and postulate on topics which they have no direct, engaged, embodied, and visceral relationship with. I have regularly observed that many professional scholars understand ecological concepts on an intellectual level yet lack primary, dirty-handed, intimate familiarity with their subjects. This can be remedied through activities like gardening and composting that are grounding, both literally and metaphorically. This firsthand awareness and mutualism with the non-human world is a salve to the mental fatigue theoreticians must otherwise endure. Furthermore, inclusion and collaboration with nonacademic communities give access to an expert knowledge not available within the confines of the ivory tower. Likewise, many sustainability activists would benefit from an expanded theoretical analysis, a luxury that the demands of advocacy work rarely permit. It is my sincere wish that, through this, the knowledge, insights, and skills of the academy be made broadly accessible to those outside of its often impermeable boundaries.

Works cited

Agyeman, J. R. (2002). Exploring the nexus: Bringing together sustainability, environmental justice and equity. *Space and Polity*, 6(1), 77–90.

Alberti, M. (2008). *Advances in urban ecology integrating humans and ecological processes in urban ecosystems*. New York: Springer Science+Business Media.

Asafu-Adjaye, J. e. (2015). *An ecomodernist manifesto*. Breakthrough Institute. Retrieved April 4, 2016, from www.ecomodernism.org

Descola, P. a. (1996). *Nature and society: Anthropological perspectives*. London: Taylor & Francis.

Elmqvist, T. e. (2013). History of urbanization and the missing ecology. In *Urbanization, biodiversity and ecosystem services: Challenges and opportunities* (pp. 13–30). Dordrecht, the Netherlands: Springer.

Gould, K. A. (2016). *Green gentrification: Urban sustainability and the struggle for environmental justice*. New York: Routledge.

Hodson, M. a. (2009). Urban ecological security: A new urban paradigm? *International Journal of Urban and Regional Research*, *31*(1), 193–215.

McClintock, N. (2010). Why farm the city? Theorizing urban agriculture through a lens of metabolic rift. *Cambridge Journal of Regions Economy and Society*, 3(2), 191–207.

McPhearson, T. e. (2016). Advancing urban ecology toward a science of cities. *BioScience*, 66(3), 198–212.

Pickett, S. T. (2016). Evolution and future of urban ecological science: Ecology in, of, and for the city. *Ecosystem Health and Sustainability*, 2(7).

Routledge, P. a. (2015). Situated solidarities and the practice of scholar-activism. *Environment and Planning D: Society and Space*, 33(3), 391–407.

Smith, A. e. (2016). *Grassroots innovation movements*. London: Taylor & Francis.

Sobel, D. (1996). *Beyond ecophobia*. Great Barrington, MA: Orion Society.
Steffen, W. e. (2015). Planetary boundaries: Guiding human development on a changing planet. *Science, 347*.
Swyngedouw, E. (2009). The antinomies of the postpolitical city: In search of a democratic politics of environmental production. *International Journal of Urban and Regional Research, 33*(3), 601–620.
Swyngedouw, E. (2010). Impossible sustainability and the post-political condition. In *Making strategies in spatial planning* (pp. 185–205). Dordrecht, the Netherlands: Springer.
Thomashow, C. e. (2015). Urban complexity and environmental education. In *Urban environmental education*. Ithaca, NY: Cornell Civic Ecology Lab. Retrieved from www.researchgate.net/profile/Angelique_Hjarding/publication/302877963_Urban_Planning_and_Environmental_Education/l

Part I

Urban ecosystem justice – a chaotic bricolage of concepts

1 Ecological alienation – roots and remedies

Nature–society dualism, complexity, nonlinearity, and the emergence of socio-ecological systems studies

An important prerequisite for understanding the origins of urban ecological alienation and of social sustainability exclusion is a familiarity with the history of nature–society dualism and the influence it has had on modernist thinking. The reconciliation of this false and destructive dualism is central to the creation of not only urban ecosystem justice but social and ecological regeneration more broadly.

The relationship between humanity and nature has been a subject of study for centuries. Conventionally framed in the past as a nature–society dualism, this dichotomy has been called "the key foundation of modernist epistemology" (Descola, 1996). Challenged by many social scientists and philosophers who believe this division between nature and culture to be false, many social theorists have proposed that humanity and nature have always been intertwined with one another, each being a coevolutionary product of the other (Figure 1.1). The more recent histories of romanticism and of frontier idealism have only further reified the mythology of a pristine wilderness separate from humanity, made possible only through the erasure of the indigenous societies that once dwelled there in relative sustainable reciprocity for centuries (Cronon, 1996). The nature–society dualism has long been reinforced by modern scientific institutions as well. This division has been reinforced by the separation of academic disciplines into the demarcated categories of the social and biophysical sciences (Capra, 2014). The reluctance of researchers on either side of the divide to cross the boundary has only furthered the disciplinary schism, leading to a "balkanization of knowledge" (McClintock, 2010). Within the biophysical sciences there is a dominant mentality of reductionism that professes to understand the world by breaking systems down into their respective subcomponents, inappropriately imposing a mechanistic worldview upon complex adaptive systems that exhibit emergent properties arising from the interactions of their many diverse parts (Levin, 1999). The rigidity of scientific determinism has contributed to a research culture that fetishizes quantification, a methodology whose limits become apparent when we consider how much in the world we choose not to measure, cannot be measured, or is unmeasurable (Eisenstein, 2018).

DOI: 10.4324/9780367858711-2

Figure 1.1 Modernity and Nature–Society Dualism
Source: photo by Elly Kellogg

Developments in the field of nonlinear mathematics in the latter part of the twentieth century, however, have given biophysical scientists an improved language and mechanisms for understanding many of the complex dynamics that characterize biological and ecological systems. These theoretical developments have created an upheaval in the ecological sciences that has resulted in a rejection of the classical view of ecosystems as static, stable, single-equilibrium phenomena. The latter paradigm has been supplanted with the recognition of ecological environments as complex, dynamic, chaotic, multi-equilibria systems. This reconceptualization has created the space for researchers to be more aware of the role that complex relationships and interactions between humans and non-humans have in shaping ecosystem form and function. Rejection of the notion of environmental forces as static and "in balance" has been embraced by a number of sociological disciplines and in turn has inspired them to abandon modernist ideas of linear social orders (Scoones, 1999). A number of socio-scientific boundary spanning disciplines have since emerged, organized around a shared interest in complex non-homeostatic systems. These include political ecology (the study of how social, political, and economic factors relate to environmental change), environmental history, ecological economics, historical ecology, ethnobiology, environmental sociology, traditional ecological knowledge, and science and technology studies. A growing recognition among even some ecologists is that human and natural systems are inherently interwoven, and that it is virtually impossible to study one in the absence of the other, leading ecology to be described as a "science of hybrid social and ecological systems" (Kueffer, 2015).

There is now a growing movement – scientific, philosophical, ecological, cultural, pedagogical, political – among activists and scholars alike that seeks to root out and challenge the deeply entrenched dualism of nature and society that is arguably responsible for driving much of the social and ecological degradation on the planet today. Looking back through history, it is evident that nature–society dualism is a fairly recent and eurocentric construct, with many instances of human societies living in "discordant harmony" with their environments (Botkin, 1992).

Anthropogenic ecosystems and anthromes

Spawning from nature–society dualism, urban ecological alienation can be in part traced to ideological and philosophical legacies that remain encoded in cultural attitudes of the present day, chiefly in the ideas that humans are separate from nature, that cities are not ecosystems, humans always have a degenerative impact on their supportive environments, and that the only way to be an environmentalist is to minimize your impact in every regard short of dying. Many people are unaware of the rich complexity of interwoven human–ecological relationships, both ancient and modern, and that history provides us with many examples of these relationships being regenerative, reciprocal, and sustained. A discussion around the existence and nature of anthropogenic ecosystems can be a starting point from which to challenge common misconceptions about human ecologies and for reframing of cities as (potentially) regenerative ecosystems.

Anthropogenic ecosystems are ecologies that have been either created by humans or heavily influenced by human activity, spanning a continuum of high-density cities to less populated "wild" environments that are shaped, tended to, or have coevolved with homo sapiens to varying degrees (Anderson, 2013).[1] Beyond an urban ecosystem justice framework, an analysis of anthropogenic ecosystems is central to nature–culture studies as so many of society's attitudes toward the earth and the non-human are informed and determined by the quality of the relationships which people have with "cultural landscapes" (Plieninger, 2012).

Understanding the varying natures of these relationships is critical for challenging the dominant Western cultural narrative that human activity inevitably results in the devastation of landscapes with unavoidable losses of ecological integrity. While this pattern may accurately characterize the trajectory of modern-day capitalist society, it is heartening to know that other human cultures have sustainably lived in the same lands for centuries while having a regenerative impact on their supportive ecosystems, at times resulting in increases in soil fertility, biodiversity, and landscape variation (Balée, 2006).[2] Disciplines of historical ecology (Crumley, 2018) and traditional ecological knowledge or TEK (Berkes, 2017) have provided considerable empirical evidence to support this assertion, while explorations of the sustainability of ancient cities may provide useful guidance to the urban predicaments of today (Barthel, 2013). Agroecology (Gliessman, 1998) and permaculture (Mollison, 1988) are two current practices that seek to re-create reciprocal human–landscape relationships that in some part draw upon the knowledge of past and present regenerative cultures.[3]

A second aspect of anthropogenic ecosystems that is crucial for understanding is that pre-industrial human societies could have significant impacts on both their local ecosystems and the global climate as well. Recent archaeological revelations have shown how the North American continent was an intensively managed agroecological system that, while unrecognizable as such to early European explorers, was capable of providing game, nuts, fruit, herbs, shelter, fungi, and fuel in amounts sufficient to sustain a population of 40–80 million indigenous people (Mann, 2005). "The great dying" (Figure 1.2) or mass die-off of 95 percent of indigenous peoples as a consequence of colonialization and exposure to European pathogens, along with the subsequent vegetative regrowth of once-cultivated land (equivalent in area to the state of California), resulted in an uptake of 2–17 billion tons of atmospheric carbon dioxide (Koch, 2019). This reduction of heat-trapping carbon dioxide in turn contributed to triggering a period of significant planetary cooling in the two centuries prior to the Industrial Revolution named "the little ice age" (Dull, 2010). Similarly, it has been postulated that ancient rice farmers increased global temperatures inadvertently through their agricultural practices. When vegetated areas are flooded (necessary for producing a rice field), organic matter breaks down anaerobically under water producing methane, a heat-trapping gas significantly more potent than carbon dioxide (Ruddiman, 2010).

Understanding the past role that humans have played in shaping the earth's climate can help to give a perspective on today's environmental predicaments if only to make us aware of the extent of anthropogenic influences on the planet and

Figure 1.2 The Great Dying – Drawn by a French Missionary of Abenaki in Maine During a Smallpox Epidemic in 1740

their potential repercussions in even pre-industrial times. It is a reminder of both systemic interconnectedness on what can often seem like a vast world and that humans and nature have a long, coevolutionary, cocreating relationship. Inasmuch as this can help us to frame even highly anthropogenic spaces, such as cities as less un-natural, it is a starting point for opening conversations about the potential for just and regenerative anthroecologies. While it is true that the activities of early cultures could have wide-ranging effects on climatic processes, care must be taken that this fact is not used discursively to condone or justify continuing negligent abuse of the earth's climate or its haphazard manipulation through irresponsible geoengineering practices (Hamilton, 2013). Ecological modernists have already attempted this, arguing that since humans have always had environmental impacts there is nothing misguided about technologically manipulating climatic processes, concentrating humans into bubbled cities, or scaling up nuclear energy without regard for matters of power or politics (Crist, 2016).

Closely related to the idea of anthropogenic ecosystems is the concept of "anthromes" as developed by geographer Erle Ellis (Ellis, 2010). A portmanteau of anthropological and biome, anthromes are intended to reflect the reality that 75 percent of the ice-free land on the planet is under some form of use or cultivation and that traditional maps of planetary biomes (tundra, desert, rainforest, etc.) are more accurately representative of the world as it was, rather than as it is. The anthromes map consists of a display of twenty-one anthropogenic environments placed on a continuum accordingly to the extent of human use occurring there. They range from relatively undisturbed "wild lands" at one end of the spectrum to dense urban settlements at the other. In between exist a variety of other cultivated spaces including rangelands, croplands, and villages. These delineations layer fairly easily on top of the "zones" concept of permaculture, a system used to designate areas of a landscape for relatively intensive or extensive appropriate uses according to their distance from "home" (Mollison, 1988). As such, anthromes maps can be helpful for conceptualizing the world according to the intensity of its anthropogenic influence and as such may be useful for making decisions regarding biodiversity planning, agriculture, agroforestry, and so on.

On the surface, the anthromes concept seems like a straightforward enough representation of anthropogenic activity that is presented in a politically neutral format. Attention must be paid, though, to how the concept could be used for either progressive or regressive purposes. Anthromes could potentially be used to justify the continued exploitation and degeneration of anthropogenic environments, with an argument reasoning "well, since these environments are already affected by humans and are therefore unnatural, what is the sense in trying to protect them?" What the anthromes map cannot account for is the quality of the relationships that humans have with the ecosystems they are in: regenerative or degenerative, reciprocal or parasitic, egalitarian or autocratic. In the absence of a more nuanced analysis, the potential for abuse is latent and high. Additionally, Ellis's concept of "Anthroecological succession", a term referring to the processes by which humans transform anthrome environments over time and affect the non-human species living there, is fraught with themes of capitalistic determinism and essentialism. By charting a course for the process of anthroecological succession from hunter-gatherer to agricultural to industrial societies, it is implied that modern industrialized humans represent a "climax state" of this process of succession, invoking largely discounted theories of "equilibrium ecology" that have since been replaced with nonequilibrium models (Scoones, 1999).

The Anthropocene

The Anthropocene has been proposed as a term to mark the beginning of a new geological epoch, one that is characterized by the enormous impact that humanity has had upon the planet and its biogeochemical processes. The combined effects on global processes resulting from climate change, urbanization, environmental degradation, biodiversity loss, nuclear experimentation, and the mass manufacturing of toxins have led many to conclude that humanity is a primary planet-shaping force. First put forth by atmospheric chemist Paul Crutzen in 2002 (Lorimer, 2012), the notion of the Anthropocene has caught hold in both the public mind, despite not being an officially accepted scientific premise.

The Anthropocene concept (along with the related idea of the "great acceleration" (McNeill, 2016) has gained widespread use in academia, the media, and popular culture with little understanding of its origins or deeper implications, posing a challenge to classical philosophical traditions, calling into question fundamental beliefs about the relationship between humanity and nature and arguably even the Cartesian "self-other" distinction. Bruno Latour describes the Anthropocene as "the most decisive philosophical, religious, anthropological and . . . political concept yet produced as an alternative to the very notions of 'modern' and 'modernity'" (Latour, 2013). Many of the most interesting conversations spawned by the Anthropocene debate are not about the Anthropocene per se but of derivative terms including "Capitalocene" (Altvater, 2016), "Cthulhucene" (Haraway, 2015), "Anthro-obscene" (Ernstson H. a., 2018), and "Plantationocene" (Davis J. e., 2019).

Anthropocene studies are key components in the urban ecosystem justice concept. Many of the central questions – humanity's relationship with nature, the role of science and technology, questions of justice, equity, power, and privilege – manifest similarly in both ideas. To these ends, the Anthropocene can serve as a parallel or complementary framing device for urban ecosystem justice studies. In some regards, the Anthropocene supports and furthers the environmental justice movement's assertation that definitions of "the environment" be expanded to include human and social processes and as such can challenge the disciplinary schism of the natural and social sciences (Houston, 2013) and false dichotomy of social and ecological activism. Unlike environmental justice, however, the Anthropocene has no explicit analysis of the socioeconomic patterns of inequitable exposure to environmental risks. Furthermore, the Anthropocene itself is neither inherently progressive nor regressive and as such has the potential to be used to further either liberatory or dominating ends.

Due to this ambiguity, and proportionate to its growing mainstream use and adoption, the Anthropocene has garnered many critics who question its implications and blind acceptance. Prominent among them is political scientist Jeremy Baskin, who argues that the Anthropocene represents an ideology and a paradigm more than it does a "neutral characteristic of a new geological epoch (providing an) ideational underpinning for a particular view of the world, which it, in turn helps to legitimate" (Baskin, 2015). Baskin's analysis consists primarily of four components that are worth summarizing concisely here. They represent many of the core elements of others' critiques that should be taken into consideration when discussing the Anthropocene and its relationship with urban ecosystem justice.

The first is that the Anthropocene "normalizes and universalizes" a certain group of humans as being the "humans" of the Anthropocene, failing to distinguish between different humans on either a spatial or temporal scale. What is implicit yet unspoken is that the humans of the Anthropocene as a whole represent wealthy citizens of Global North nations and largely exclude residents of the Global South. The former is driving many of the human activities that characterize the Anthropocene and have benefited more from the material gains resulting from the planetary exploitation. Further, by naming the Anthropocene as such it is given a biological definition. This conveniently avoids the problem of referencing the human epoch as "capitalism" or "neoliberalism" and the normative implications either term would carry.

Second, Baskin sees the Anthropocene as being philosophically problematic for the relationship between humans and nature. While arguably, the Anthropocene functions to mend the rift between humans and non-humans, nature and culture, it does so only to elevate humanity over nature, reaffirming our dominance over the non-human world. While it is no longer possible to see ourselves as something separate from nature, the Anthropocene allows us to recognize the extent of our interwovenness while still retaining mastery. In this regard, Baskin sees the Anthropocene concept as a product of modernism. Its proponents reject prevailing beliefs in the ecological sciences that call for "multi-natural" approaches to nature that fall somewhere in between harmony and mastery. Rather, the Anthropocene portrays nature as a "colonial subject" of humanity.

Baskin believes that it is important to respect the otherness, exteriorness, and agency of nature, seeing "artifice and wilderness" as only two ends of a long spectrum, with much in the middle. To typify this, he quotes Kenneth Worthy as stating "[n]ature in a rainforest or a coral reef, even with the marks of humanity, reveals elements of complexity and dynamism not seen in plastic pink flamingoes and jumbo jets" (Worthy, 2013).

Baskin is also critical of Anthropocene proponents' relatively uncritical acceptance of technology. He refers to a trinity of techniques called upon for the global management of planetary systems. They are clear management of earth systems, guidance by experts, and the use of advanced technology. The prevailing attitude among this group is that while we may be facing dire issues, such as climate change, a technological solution will present itself that will remedy the problem. Many of these proposed technologies are global in scale and could be described as forms of geoengineering including oceanic fertilization and atmospheric sulfate injection.

All the concerns that Baskin raises could be equally applied to urban ecosystem justice. Within reframing the relationship between humans and nature, especially in cities, we must still respect and appreciate the "otherness" and complexity of nonanthropogenic ecosystems, going so far as to maintain an "ecology of separation" (Neyrat, 2018) that does not grant humans mastery over nature in the process of their reunification. Furthermore, urban ecosystem justice activists must have a wariness about the overreliance on sophisticated technologies and technofixes (Ruser, 2016) and a developed analysis of equity, fairness, and intersectionality that recognizes the primary drivers of injustice and planetary degradation. As increasingly there is a push for the widescale adoption of "smart cities" technologies that seek to monitor and quantify every conceivable dimension of a city's metabolism (air quality, greenhouse gases, traffic, energy usage, population fluxes, toilet flushes, etc.), it's important to remain critically skeptical of these technological incursions and continue to ask for whom this technology is intended and to be vigilant against potential undemocratic misuses (Colding, 2019). These dystopic visions are manifesting in the emerging discourse of ecomodernism as it pertains to the Anthropocene.

Ecomodernism and the "Great Anthropocene"

Proponents of the Anthropocene concept include the Breakthrough Institute, a pro-nuclear, anti-green think tank founded in 2003 by Ted Nordhaus and Michael Schellenberger that is focused on promoting solutions to climate change driven through capitalist technological investment (Angus, 2015). The Breakthrough Institute advocates for "ecological modernization", a philosophy that posits that economic growth and technological expansion are both ecologically compatible and can be used as tools to promote environmental reforms (York, 2003). Summarized in a popular format through the publication of the "Ecomodernist Manifesto" (Shellenberger, 2015), the authors advocate for avoiding climate change entirely through technofix solutions, believing any environmental obstacle may be overcome through technological and economic means. Rather than denying climate change, the manifesto calls for expansion of nuclear and natural gas energy production, further urbanization of the world's population, as well as intensification of food production via genetic modification of crops and indoor vertical farming systems. Through a combination of these activities, the remaining land on the planet's surface can be used for biodiversity conservation and the regrowth of forest which would result in a reduction of atmospheric carbon dioxide.

The ecomodernists (as the manifesto's authors refer to themselves) have additionally seized upon the idea of the Anthropocene. Rather than seeing it as a gloom and doom scenario, they propose that rather it is an opportunity to raise the standards of living of all people and that a "good" or even "great" Anthropocene would be possible if societies would reject the idea that harmonizing with nature is necessary in order to avert ecological collapse. In this sense, they put forth that because human exploitation of nature has only yielded greater wealth and standards of living for people then increased modernization will only achieve better results. Absent from their argument is any consideration of planetary boundaries or tipping points or that relative global ecosystem stability may pass critical thresholds before their environmentally stable techno-utopia may be obtained. Particularly insidious is the ecomodernists offer to relieve their adherents of any environmental guilt or culpability, claiming that activities such as consumption and anthropogenically influenced environments are ingrained within human nature and have been going on for our entire history. Immediately, humanity's "original environmental sin" is forgiven, and a path forward is laid out that rewards its followers with greater material affluence. Concerns regarding equality and justice are allayed through assurances that all global residents will benefit as the collective tide rises. Nowhere, however, is any consideration given to the idea of responsiveness to feedback. Early human societies certainly did have heavy influences on their environments, yet had to do so within limits as exceeding the carrying capacity of their supportive ecosystem would result in negative feedback loops on their populations. Only through global capitalism has it been possible to export negative externalities and extract resources on the extent of a global scale as to lengthen the feedback loops so as to be willfully ignorant of them. By pushing the technological envelope, the ecomodernists make a dangerous gamble.

Another Anthropocene is possible

From a combination of critiques of the Anthropocene and suggestions for alternative Anthropocene realizations, it seems plausible that "another Anthropocene is possible".

In order for that vision to be realized, progressive scholars from both sides of the natural/social disciplinary fence will need to cocreate an alternative vision that incorporates the principles of sustainable ecological design, community-based resilience, a respect for the diversity of people and their lived experiences, and a political space that is open and tolerant of debate and dissenting viewpoints. While I agree that a state of planetary crisis does exist, we cannot allow fear to silence differing viewpoints and to force a blind acceptance of dangerous, centralized, and non-democratic technological solutions. An entire toolbox of solutions is needed, one which contains enough possibilities to be tailor-made in accordance to the needs, knowledge, and experience of the diverse communities on the planet. Urgent action must be taken, but it must be as open and as democratic as possible so as not to result in the implementation of a regime of global technocratic totalitarian control.

In these regards, the debates over justice in the Anthropocene closely mirror those in urban ecosystem justice and in just urban transitions overall (Hughes, 2020). Cities must be planned in a manner that is radically democratic and participatory, centering issues of justice and equity along with the voices of the most marginalized populations. While technology can certainly play a role, it should remain a tool of the people and not the primary milieu through which we experience reality nor a means through which technocrats may pursue dominance. The relationship between humans and non-humans in cities must be based on an ethic of reciprocity that seeks to unify the nature–culture divide – humans cannot wantonly degrade non-humans but neither should they be separated from them by glass enclosures. Doing so will create a populace devoid of ecological love and rife with environmental anxiety.

Urban ecology

What is an ecosystem? In response to this question, many people would imagine rainforests, coral reefs, African savannahs, or another exotic environment. Ecosystems are commonly thought to have abundant animal and plant life, clean air, pure water, and to be otherwise "healthy". In few of these picturesque scenes would there be any mention of humans or human activity outside of a romanticized primitive state.

Cities then, as constructed environments with very high concentrations of humans, would surely be considered the antitheses of ecosystems. Filthy, noisy, polluted, degraded, and devoid of species diversity, these *Homo sapiens* monocultures would historically be considered by few to be ecological entities. An ecosystem, however, is defined as being a community of organisms interacting with the living and nonliving elements of their environment – all criteria that a city meets. Nothing says that an ecosystem must exclude humans or that it must exceed any threshold of "purity".

In cities, people engage in relationships with the many other living creatures and nonliving buildings, sidewalks, soils, parking lots, waterways, community gardens, fountains, factories, and atmosphere surrounding them, mixing together to create a beautiful, complex, and dynamic mosaic of life. In addition to being places of culture, industry, entertainment, art, policy making, finance, religion, cities are also ecological spaces. They are, as described by urban designer Marinna Alberti, "hybrid systems emerging from interactions between human and ecological systems" (Alberti, 2003) – anthropogenic ecosystems in the most extreme sense. This recognition has given rise to the field of urban ecology as a legitimate discipline with its own self-named journal. Urban ecology research is now highly interdisciplinary, commonly combining both biophysical sciences and sociocultural studies to form its conceptual approaches (Childers, 2014). In this section, I will explore urban ecology as a socio-ecological framework that is integral to urban ecosystem justice. I will explore its history, its relationship with other disciplines, and a few of the key concepts of urban ecology and how they relate to the field of study.

Traditionally, ecologists have shunned cities as objects of study. In addition to the perception of human-dominated spaces as being unnatural and therefore non-ecological, cities are also highly complex systems with ambiguous boundaries. As ecologists have preferred to work with simple models with a minimum of variables and well-defined limits such as ponds, small forests, or islands, cities made nonideal subjects for study. This attitude has been reinforced by much classical environmental thinking where nature is thought of as the world outside of human creation. The excluded social environment was regarded as "the scene of humans interacting only with each other in the absence of nature" (Worster, 1988). There have been, however, parallel strains of thinking that have at least recognized a naturalness of cities, if not falling short of explicitly describing them as ecosystems.

"Organic theory" is one example dating back to the late eighteenth century that was dominant theory among middle-class urban scholars. It analogously compared the structure and function of a city to a human body, with streets representing arteries and so on. It was favored by the ruling classes who imposed a form of social Darwinism onto the theory that justified their roles as the "brains" or "fists" of the city organism (Melosi, 2003). Due to its limited use as a metaphor and elitist overtones, the organic theory is seldom invoked today despite its rough description of certain urban ecological processes. A more recent example from the 1920s is that of the Chicago School of Sociology. The Chicago School attempted to develop a general theory of urban social systems employing several ecological concepts including succession and competition. Like organic theory, though, it largely employed analogies rather than mechanistic models. The Chicago School was additionally flawed in that it failed to incorporate the contemporary ecological thinking of its time – already by then the notion of ecological succession and climax communities was being called into question. The "naturalization" of successional processes to describe one social group displacing another could easily be used as a justification for the gentrification of marginalized communities. Furthermore, the Chicago School idealized rural life and saw cities as fundamentally negative and blighted, an attitude that has carried over into environmental philosophy today (Grove, 2015). Over time, the influence of the Chicago School has faded and has been replaced by urban sociological models that employ sociocultural phenomena over ecological analogies. While certainly preferable, these frameworks nonetheless effectively separated the social from the ecological in the study of cities, a rift that has only recently begun to be mended.

Starting in the years following World War II, recognizing that human activity and urbanization were having significant impacts on ecosystems, certain ecologists began looking at the ecological processes occurring within and around cities. For the most part, these early urban ecologists were importing their knowledge and skill set from less disturbed environments and comparing them to analogous urban environments such as woodlands, meadows, cemeteries, streams, wetlands, etc., with the assumed drivers of change being anthropogenic (Pickett, 2016). Studies of urban wildlife, invasive species, water body impacts, and forest transformation dominated the topics of study, with most concluding that human impact had a negative impact on biodiversity and ecosystem health overall.

While this period of study focuses on standard ecological functions and structures inside an urban environment and on "remnant ecology" or the non-human ecological processes taking place inside of cities, since the late 1990s urban ecology has now shifted to a more holistic approach where the city itself is studied as an ecosystem with humans as an integral component (Grove, 2015). This paradigmatic shift undergone could be described as a movement from an analysis of *ecology in cities* to the study of the *ecology of cities*. Partly due to the influence of nonlinear thinking, urban ecology now considers how social factors affect urban processes, and has synthesized its analysis with social science disciplines such as economics, urban design, political ecology, complexity science, and geography (McPhearson, 2016). Urban ecology now looks at social and biophysical processes as "pervasive, reciprocal, and intertwined" (Pickett, 2016) with the causes for change residing at the interface of their common feedbacks. Along with this, social norms and power relationships are questioned, with urban dynamics changing over spatial and temporal dimensions. These parallels of social sciences and ecological sciences have created an integrated socio-ecological systems science of urban ecology. Ecology of the city now frames urban environments as complex adaptive socio-ecological systems, the result of coevolutionary processes between the human and non-human aspects of the urban ecosystem (Alberti, 2003). As stated by urban scholar Marina Alberti, "an ecology that does not involve humans in its theories and experiments will rapidly evolve into paleoecology".

Although the two paradigms of *in* and *of* have different areas of focus, it should be noted that the latter has not overthrown the former in a Kuhnian sense but rather has expanded upon it while still incorporating many of its techniques. Empirical studies of ecological populations and the effects of human impacts are still studied, just with an increased awareness of political and social dynamics and change over time.

Most recently, yet another paradigm shift has occurred in the field of urban ecology that is best described as an *ecology for cities* (McPhearson, 2016); (Pickett, 2016). *Ecology for the city* could be described as a way of extending the land ethic of Aldo Leopold to the urban environment, where a sense of concern and community is extended beyond humans to include entire ecological communities of soils, water, animals, and plants (Cadenasso, 2015). Furthermore, drawing from the insights of both ecology "*in*" and "*of*" the city paradigms, ecology *for* the city advocates taking the integrated research of urban ecology beyond scholarly and research disciplines and adding an ethical component to scientific understanding.. Operationalized *ecology for the city* calls for engaging in dialogue with citizens, groups, governmental agencies, and technical experts to jointly identify research goals and collaboratively produce useful data. Reciprocally, it is important for each of these groups themselves to have a good understanding of urban ecological science. Overall, in the *for* paradigm differences in power, privilege, and wealth are considered and given respect to. In these senses, *ecology for the city* takes the technical understandings of *in* and *of* and applies them to the normative goals of achieving social justice, economic equity, and environmental well-being (Pickett, 2016).

The *ecology for the city* paradigm so far shows the greatest promise for urban ecology to be in more direct communication with the environmental justice movement. These are two fields that share much in common in regard to their topics and spatial region of study yet to date have remained largely separated (Boone, 2012). Urban ecology has historically resisted making normative statements, preferring to maintain the appearance of scientific objectivity. Environmental justice, on the other hand, while having operated as a powerful tool for activists has been critiqued for largely omitting the "environment" (biophysical factors), with most studies being focused on human impacts (Boone, 2012). It is furthermore suggested that environmental justice activists would benefit by incorporating a more systems-oriented, future normative perspective utilized by sustainability researchers in order to move beyond the purely reactive "militant particularism" (Harvey, 2008) that characterizes the movement. A potentially powerful synergy between these two disciplines could aggregate around the *ecology for the city* concept and idea that additionally shares much in common with the Urban Ecosystem Justice framework.

Novel ecosystems

An important and central concept to urban ecology is that of the novel ecosystem. A novel or emergent ecosystem is defined as a new and unplanned and unmanaged arrangement of organisms resulting from human activity that has no precedent or analogue in nature. They are further explained as being a "system of abiotic, biotic and social components (and their interactions) that, by virtue of human influence, differ from those that prevailed historically, having a tendency to self-organize and manifest novel qualities without intensive human management" (Hobbs, 2013). It is assumed that there is little to no possibility of restoring a novel ecosystem to its original, nonnovel state (those types of ecosystems where this possibility remains are referred to as "hybrid"). Novel ecosystems cover an estimated 35 percent of the world's ice-free land (Marris, 2013) and many, if not most, urban landscapes are dominated by novel ecosystems and processes. Novel ecosystem studies are, for this reason, a central topic to urban ecology. The discussion surrounding them is surprisingly political and bears many similarities to that being held around the Anthropocene.

Traditionally, ecologists generally avoid novel ecosystems as subjects of study as they frequently possess considerably less biodiversity and systems complexity than nonnovel ecosystems. As thousands of years of coevolutionary processes are required to create the complex relationships between animals, plants, fungi, bacteria, soils, water, and other living and nonliving components found in a high-functioning ecosystem, recently emerging novel systems are likely to possess less complexity. Furthermore, many novel ecosystems are dominated by a handful of species that are considered invasive and pose a threat to surrounding environments that are populated by more native species.

There are many criticisms of the novel ecosystems concept, primarily from the community of conservation and restoration biologists. Chiefly among them are that the novel ecosystem label may provide a "license to trash" for corporations looking to accelerate approval processes for development plans or avoid paying for environmental restoration (Murcia, 2014). By devaluing ecosystems as hopelessly novel, opponents of the concept fear it may legitimize the tendency to ignore the long-term impacts of development, pollution, and depletion of natural resources. Furthermore, they question the validity of the idea of thresholds, tipping points, and irreversibility of ecosystems in the gradient of historic to novelty.

Other conservationists, however, counter that the concept of novel ecosystems alone does not give license to those who would disrupt or modify ecosystems, as that has been occurring for much of human history. Rather, they see utility in the concept for making decisions about the likelihood of success in different interventions and in setting priorities. Interestingly, they also point out that originally preservationists had the similar concerns about the idea of restoration ecology, fearful that it would devalue their efforts (Hobbs, 2014).

Recognizing that there are poorly developed guiding principles in place for making decisions and judgments about novel ecosystems, political ecologist Paul Robbins has identified the need for "evolving a political ethics for novel ecologies". Traditional ecological ethics, he points out, have little to say in regard to novel ecologies on account of the fact that their focus has been on "introducing non-anthropogenic principles to extend care and value to non-humans in, or and for themselves". Robbins calls for a more engaged political ethics that would grapple with the complexity of the ecological, social, and political underpinnings of decision-making. Contrary to rejecting the contribution of Edenic scientists, however, Robbins sees conservation, restorations, and invasion biologists as being leaders in the decision-making process on account of their knowledge of interspecific relationships, succession, tipping points, and complexity (Robbins, 2014).

Yet another argument is that studying and even embracing novel ecosystems have value, particularly in urban environments. Coupled with the recognition of urban ecology as a respected discipline and the acceptance of the Anthropocene concept, the study of novel ecosystems (and the related field, disturbance biology (Walker L. R., 2012)) has grown into a legitimate field of study. While many novel urban ecosystems may consist of noxious and invasive species, an understanding that novel ecosystems may nonetheless provide many essential ecosystem services, with carbon sequestration, pollution reduction, food production, water purification, and habitat creation among them, is emerging. Seen through this lens, the disturbance tolerant, "ruderal" (Stoetzer, 2018), or "blasted" (Tsing, 2014) plants and animals that comprise these novel urban ecosystems could be thought of as being "native" or "indigenous" to the urban environment. While native plant and animal assemblages once existed on the land where cities now stand, massive abiotic changes have since taken place. These original communities would not survive today without re-creation of the original soil and nutrient conditions. For this reason, any attempt to restore pre-urban native ecologies to today's urban environments would simply be, as put by Harvard professor Peter Del Tredici, "gardening dressed up as ecology" (Del Tredici, 2010). This point of view is consistent with the nonequilibrium ecological paradigm that recognizes that ecosystems are in a continual state of change and transformation. While it is not meant to undermine the efforts of those seeking to protect large and relatively undisturbed areas of habitat, it does call into question the wisdom of expending considerable resources to "restore" a vastly transformed ecosystem back to a particular, idealized, historical state. This outlook is concomitant with the emerging discourse of "critical invasion science" that questions the assumptions and orthodoxy of invasion biology, believing it inadequately addresses the complex intersectionality of plants, animals, humans, science, history, power, politics, and culture (Kull, 2018).

From a perspective of youth urban environmental education in an urban ecosystem justice context, I find great value in discussing the novel ecosystems concept. I believe it is important for youth to think of nature as not something external that exists only outside of cities, but rather something that can be found within cities as well. By studying and appreciating the functioning of novel urban ecosystems and the environmental services they provide, youth can begin to develop an understanding and appreciation of ecological processes. From there, it is possible to discuss the differences and contrasts between historic and novel systems and to appreciate the reasons why particular historic systems may need preservation. In this sense, novel ecosystems can help to provide "cultural ecosystem services" (Collier, 2014) and form the basis of a "science of transition" that could mainstream sustainability sciences and create normative social goals (Collier, 2015). Furthermore, the novel ecosystems concept provides a framework for discussing the nuances of what get called "invasive species" as well as their relative merits, learning about the role that these plants and animals can play in building urban ecosystem justice despite their non-native status (Frawley, 2014).

In conclusion, studying novel ecosystems can help to challenge urban–rural dichotomies and the fundamentalism of the "Edenic" sciences. At the same time, it is important to be wary of corporate interests appropriating novel ecosystems discourse as a pretext for unrestrained development, habitat destruction, and landscape degradation. By claiming "all ecosystems are already novel, or have already exceeded the threshold", they may seek to discredit the work of those attempting to preserve biodiversity on a global scale.

Urban metabolism

Urban metabolism is a concept used in a variety of fields to model the various inputs and outputs of material and energetic resources going through a city and how they are transformed into the built environment, biomass, and waste (Decker, 2000). Metabolic framings of cities are used in disciplines as diverse as urban ecology, political ecology, industrial ecology, ecological economics, and political economy and have a history of use extending back to the nineteenth century (McClintock, 2010). Applications of this concept range from the purely materialistic modeling of resource exchanges to a tool for studying inequalities of resource distribution in neoliberal economies (Broto, 2012). While urban metabolism has its limitations in being able to describe the intricacies of the complex adaptive processes of socio-ecological systems, and while its use as a biological metaphor (Figure 1.3) may be an inappropriate characterization of "cities as organisms" (Golubiewski, 2012), I believe that it still has considerable value as a general framework for understanding many of the ecological, social, and economic processes that shape urban environments. As such, it is a key concept in the Urban Ecosystem Justice platform. Before delving more deeply into practical applications for the urban metabolism concept, it is worthwhile to explore its history and use in theory.

Geographer/urban political ecologist Eric Swyngedouw frequently employs the metabolism metaphor to describe societal–natural processes occurring within cities (Swyngedouw, 2006). Metabolism, as framed by Swyngedouw, is useful as a concept for describing the flows of resources that act as the inputs (food, water, building materials, energy, labor, etc.) and outputs (waste) from an urban environment. Tracing the term back to its original scientific meaning of being the "totality of the exchanges of energy and substances between an organism and its environment", Swyngedouw explains how the concept of metabolism was subsequently used by Marx to describe the flow of resources, labor, and material from rural to urban environments (according to Swyngedouw, Marx and Engels were the first to use the concept of metabolism to describe the dynamics of socio-environmental evolution). Marx saw this metabolic flow as resulting in the degradation of soils, environments, and people and the accumulation of garbage and sewage in the city. This process resulted in what Marx refers to as "metabolic rift" – or the alienation and separation of society from nature and people from land. (Interestingly, though, Swyngedouw proposes that by focusing on labor as purely a social process, modern Marxist philosophers have perpetuated and contributed to the original metabolic rift. He decries what he sees as a tendency in twentieth-century socialist traditions to see Marxism as a purely sociological theory that has created a "partial blindness in the twentieth-century social sciences to questions of political ecology and socio-ecological metabolisms") (Swyngedouw, 2006).

Swyngedouw contributes a unique analysis of urban metabolism through the study of the political ecology of urbanizations, applying a lens of justice and equality that is key to an urban ecosystem justice outlook. According to Swyngedouw, as cities expand, they create new socio-natural systems within themselves and transform the surrounding rural lands as well. This process is not an egalitarian one, as it creates a disparity of wealth and poverty over a broad geographic range. Questioning and analyzing these power relations are critical to Swyngedouw:

> [T]there is no such thing as an unsustainable city in general; rather, there are a series of urban and environmental processes that harm some social groups while benefitting others. It follows that a just urban socio-environmental perspective must always consider who gains and who pays and ask serious questions about the multiple power relations – and the networked and scalar geometries of those relations – whereby deeply unjust metabolic processes are produced and maintained.

In this way, Swyngedouw is stating that the struggle to bring about just urban environments is closely linked with the struggle against exploitative socioeconomic relations. His critique stands out among urban metabolism scholars in that it considers issues of access and equity as core components where many others employ metabolic modeling for purely technocratic purposes.

In this regard, socio-ecological metabolic processes are necessarily political and integral to any social project. Likewise, any political project is also an ecological one. The broader social justice potential of urban metabolism has been questioned by asking if Lefebvre's notion of "the right to the city" necessarily implies a "right to metabolism" (Heynen, 2006). The question of whether urban metabolism contains emancipatory potential is yet to be fully answered or adequately explored,

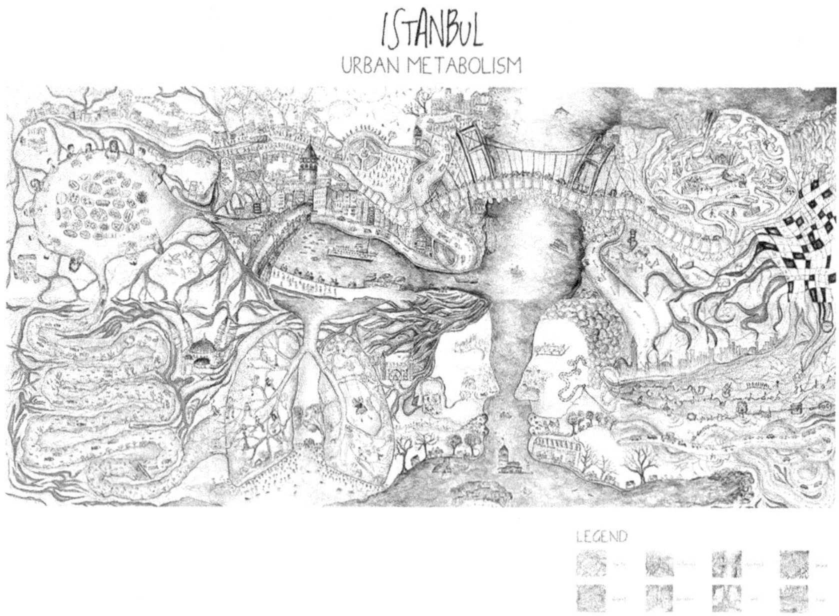

Figure 1.3 Urban Metabolism – Art by Deniz Ayaz

although there have been some preliminary explorations of the idea of a decentralized and user-driven urban metabolism (Chelleri, 2016).

Ecological metabolisms

The urban metabolism framework, as well as its conceptual depictions of material flows, closely parallels ideas regarding metabolisms in the ecology. In the ecological sciences, ecosystems can be placed very generally into two separate categories accordingly to the nature of their metabolic processes. The first category is described as being heterotrophic or linear ecosystems. Examples of non-anthropogenic linear ecosystems include streams, whose leaf-litter eating invertebrate residents depend on leaves to fall into the stream and for their wastes to be carried away by the flow of the water. The typical modern city is also an example of a linear ecosystem. Flow-paths of energetic and material resources from around the globe make their way into urban centers where they are consumed, producing significant quantities of often toxic waste by-products that are exported to other areas (although many of these toxins have continued to accumulate in the soils and waters of cities themselves) (Girardet, 1992). The metabolic processes of human societies grow increasingly linear as neoliberal economies progress, commonly siphoning resources from the Global South into wealthy cities of Global North nations.

The second category of ecosystems are referred to as being autotrophic (self-feeding) or cyclical. Examples of autotrophic ecosystems include ponds, which generate much of their own food through algal photosynthesis. The waste products of the organisms that eat the algae are then processed by bacteria that are part of the system. While there certainly still inputs and outputs into an autotrophic system (beavers coming in, fish being caught and removed), relatively speaking they are self-contained (Girardet, 1992).

When imagining ways for cities to be more sustainable, it can be helpful to think of structuring them to function cyclically or more similarly to an autotrophic system. Such a city would be capable of producing a greater amount of its food, water, energy, and building materials internally, as well as having the capacity to process its own wastes. A city with a cyclical metabolism would very likely have less of impact, environmentally speaking, on the rest of the planet (Girardet, 2008). By producing resources locally, the carbon footprint requirement of transporting goods is reduced. Residents then, on account of living in close proximity to the origins of their resources, will be more aware of the impact that their lifestyle and consumer choices are having on the local environment. Additionally, by processing wastes internally, a greater incentive to produce fewer toxic materials is produced, as they would not be externalized to distant locations.

It is important to point out, however, that just as Eric Swyngedouw emphasizes, an ecologically cyclical city is not guaranteed to be socially just and equitable (Swyngedouw, 2006). The practice of urban ecological securitization explicitly attempts to build cities with circular metabolisms (Hodson, 2009). These cities are designed to be providing their residents with close to all of their material needs, while safeguarding them from any conceivable ecological risk, such as natural disasters, drought, famine, or climate change. Such closed-loop systems create a hermetically sealed environment disconnected and independent of the surrounding landscape, similar to the vision espoused by ecomodernism (Asafu-Adjaye, 2015). They are also commonly excusive enclaves of wealth, affordable only to the most privileged. Poor residents are displaced to the urban periphery where they are subject to the brunt of natural disturbances and resource shortages, victims of an ecological gentrification. Such patterns are already seen today, particularly in Global South megacities where the urban poor are relegated to slums constructed on hazardous terrain, exposed to the dangers of floods, fires, hurricanes, landslides, and pollution (Davis M., 2007). Similarly, a common trope of techno-sustainability regimes is to invoke "closed-loop" imagery when describing ecologically exclusionary systems. This language is misleading in a number of regards, one being that completely "closed-loop" systems do not exist, ecologically speaking. Boundaries are made around ecosystems for the purposes of simplicity and convenience when modeling, however, in truth no impermeable boundaries exist around any ecosystem – all ecosystems overlap and exchange with one another, either through nutrient cycling, movement of organisms, energy exchange, etc. Even the earth itself is not a closed system – energy goes in and out and meteorites routinely add to the planet's mass. Furthermore, food growing or waste recycling systems that are billed as being closed-loop commonly are built in technologically sophisticated systems that require high start-up and operational energetic and monetary costs. Such complex and tightly coupled systems are subject to cascading failures (Perrow, 2011) and are limited in their broad-scale applicability on account of their cost and technical demands. Closed-loop systems may be necessary in the environment of a spacecraft, but their development on the earth is a refutation of the inherent interconnectedness central to ecological thinking.

Urban ecosystem justice, therefore, must strike a balance between the two metabolic extremes of linear, hyper-capitalistic machines of consumption and the exclusive "closed-loop" ecological enclaves formed of security culture. This can be thought of as a type of "just" hybrid metabolism – one that produces a significant amount of its needs internally while still maintaining a dependence on external resources, albeit derived from local regions and extracted in a sustainable, reciprocal, and perhaps even reflexive manner. In this regard, it is desirable for just urban metabolisms to become more cyclical in nature, however, not going so far as to become an exclusive enclave of ecological security. In order to achieve this, two points must be emphasized:

> The first is that a metabolically just city must maintain a healthy and reciprocal relationship with the surrounding peri-urban and ex-urban landscape. For instance, increased capacity for urban agriculture and food production is an important goal for a just city. Local food production can help to increase urban resident's access to nutritious foods, and can create sustainable employment, as well. Due to ratios of agricultural space to population, however, it is unlikely that any city in the near future would be capable of producing the majority of the caloric requirement for its population within its boundaries (Martellozzo, 2014). For this reason, cities will remain dependent on farms and water bodies in their near vicinities for providing them with food. This relationship would ideally foster a close interdependence and interchange between rural and urban spaces, with cities acting as centers of cultural, social, and economic activities that will enrich the lives of rural residents. By being within close proximity to a city, environmental impacts on landscapes will be more visible, making more difficult to "externalize" environmental impacts to the "ecological hinterlands" of invisible and distant regions in the Global South, as is done conventionally (Monstadt, 2009). In this sense, smaller cities with a reduced urban to rural gradient may be inherently more sensitive to feedback loops from surrounding land bases. From there, it may be possible to begin laying the basis for building progressive solidarity between oppressed rural and urban populations.

Furthermore, in ecological terms, a just city would additionally strive to move beyond a paradigm of "sustainability" to one of "regeneration". In a regenerative system human needs can be met while ecosystem services such as water purity, soil fertility, and biodiversity are simultaneously increased. By-products generated by human activity are nontoxic and compostable and are used to build the health and richness of soils in the urban and peri-urban sphere. This stands in contrast to sustainability, which implies merely being able to continue on a current trajectory without it resulting in ecological degeneration, never going so far as to repair the damages caused by previous environmental harms (Cole, 2012).

It is important to point out that many cities, prior to the mid-nineteenth century, had a far more cyclical and reciprocal relationship with farms is the peri-urban sphere. It was a common practice for night soils, the manures of both humans and non-human animals, to be collected from cities and redistributed via market processes to farms in the urban periphery. In exchange, the fertilized fields of the farms would grow produce and hay for consumption by urban humans and horses. In this sense, the nutrients used for food production remained in a relatively circulatory state. By the end of the nineteenth century, this practice had largely been abandoned. Notions of sanitation and cleanliness ushered in the more widespread use of sewer systems, and increasing reliance of horse power saturated the manure market, removing the profit incentive to transport manures (the eventual replacement of horses by automobiles ultimately eliminated the horse manure market). With nutrients no longer be returned to soils, the fertility of the land declined along with the health of both farmers and factory workers, resulting in what historians refer to as "the great separation" (Atkins, 2012). What followed was cities developing a mostly parasitic relationship with the surrounding rural areas and the ushering in of the linear metabolism (Odum, 1989). The following quote by Engels captures the essence of this metabolic rift:

> Abolition of the antithesis between town and country is not merely possible. It has become a direct necessity of industrial production itself, just as it has become a necessity of agricultural production and, besides, of public health. The present poisoning of the air, water and land can be put to an end to only by the fusion of town and country; and only such fusion will change the situation of the masses now languishing in the towns, and enable their excrement to be used for the production of plants instead of for the production of disease.
> (Engels, 1959)

Few would advocate for recreating conditions of mid-nineteenth-century cities in cities of today – certainly many aspects of the sanitation movement resulted in improved public health, and accumulations of horse manure in cities presented a significant environmental dilemma (McShane, 1997). The essential practice, however, of returning nutrients to the soil has been central to the sustainability of any society. Advances in the microbiological sciences, composting technology, and public health have now made it possible to safely and efficiently compost all animal manures and food wastes, turning them into a finished product that will enhance the health of soils (Jenkins, 2019); (Redlinger, 2001). While arguably this practice continues in the form of reapplying biosolids, or septic sludge, to agricultural fields, the sewage treatment technological complex is intensive in its use of water, chemicals, and energy. Furthermore, by nature of its design it fails to segregate human wastes from industrial pollutants, producing an end product that contains toxic residues. The SOIL project based in Haiti serves as an excellent sustainable model of how human by-products can be used to generate both soil fertility and economic opportunities (SOIL, 2015). The project employs local residents to keep composting toilets clean and stocked with carbon-based sawdust. When humanure containers are full, they are brought to a centralized facility where they are composted at high temperatures, ensuring pathogen death. Finished compost is then applied to Haitian fields and forests to replace soils lost due to deforestation, itself a consequence of French colonialism (Macintyre, 2010). Barriers to the implementation of such practices in the Global North include cultural attitudes and the significant investments already made in sewer technologies.

The second criteria for just hybrid urban metabolisms is equity and access. In order to prevent cities from turning into the equivalent of environmentally closed-loop gated communities, assurances against economic displacement must be institutionalized. Quite importantly, a city cannot be thought of as a neoliberal investment machine (Molotch, 1976). Cities are communities, places where multigenerational family relationships are built and lives are led. Community will suffer when cities are thought only as financial engines, with homes and property being purchased and sold merely as commodities, and "investment properties" used to generate rental income from the poor. Too often today, city governments are managed through networks of cronyism, with lucrative development contracts treated like tradable stocks among shareholders of a corporation and the mayor acting in the equivalent role of the CEO. To invert this dynamic and to break the toxic cycle of investments that frequently hinder the growth of urban communities, it will be helpful to invoke "the right to the city" (Harvey, 2008) or the idea that citizens of a city have an inherent right to be involved in the governance and planning of their city and that their needs and interests should be given priority over those of businesses and real estate developers.

As part of this strategy, safeguards against displacement resulting from gentrification, particularly ecological gentrification, need to be put in place. The challenge lies in that many "green amenities" of cities, such as parks, forests, community gardens, public transportation, etc., commonly have the effect of increasing property values. While such amenities should be accessible and available to all urban residents, their implementation usually has the effect of increasing property values and displacing low-income urban residents (Dooling, 2009). In keeping with the spirit of the slogan "cities for people, not for profits", assurances such as land trusts, home ownership assistance programs, rent control, and public housing need to be put in place (Brenner, 2009). While these programs often offer little more than temporary solutions in today's economic climate, any conversation about sustainable cities must take economic and social factors into account as much as environmental ones.

Resilience

Resilience is a concept that has been adopted and used broadly in fields ranging from urban planning, environmental conservation, disaster preparedness, corporate management, and increasingly, private security and cybersecurity (Weil, 2020). With distinct origins and meanings in ecology, engineering, and psychology, resilience is frequently used today as a conflated mash-up of all three, being used interchangeably, or as a surrogate for, "sustainability" (Sehgal, 2015). Taking advantage of the vagueness of the term and the general confusion regarding its precise meaning, a class of "resiliency experts" have emerged that market their consultancy services to city governments and businesses, selling a highly reductive and generalizable variety of resilience that's oriented toward risk-management strategies (DHS, 2020), often with the goal of responding to forces of "disruption". Although resilience has been heavily co-opted by corporate and neoliberal interests in recent years, I still believe the idea has potential value (particularly in its ecological context) and that it is important to discuss on account of its prevalence in critical urban studies. A boundary spanning socio-ecological construct, resilience has many social, environmental, and political implications that make it integral to an urban ecosystem justice analysis.

Like sustainability, resilience can be used as a metaphor containing important social and political meanings. Following its vernacular definition, resilience can describe the ability of an individual or a system to "bounce back" following a disturbance or perturbation. The image of the "resilient city" is often used to describe an urban area that is capable of maintaining function and being adaptable in the face of change. On top of its use as a metaphor, resilience also has a clearly defined meaning within the ecological sciences (Pickett et al., 2014). In order to employ resilience as an operative tool for the design of sustainable, healthy, and egalitarian cities, it is important to have a thorough and in-depth understanding of its meaning as a scientific concept.

Developed by C.S. Holling in the 1970s, ecological resilience challenged the previously held notion in empirical ecology (influenced by physics and mathematics) that ecosystems reached "climax" or equilibrium states. Resilience theory postulated that instead, ecosystems are constantly in a state of flux, moving from states of increasingly concentrated energy and resources to a state of disorganization and release of stored capital. Following this release of resources, the system then goes through a subsequent period of reorganization and resource accumulation. In resilience theory, this process is referred to as the "adaptive cycle" (Figure 1.4). In this regard, no singular "optimal state" for any ecosystem exists but rather a cycle of phases in which the system can function and exist within in the face of disturbance and change (Walker B. a., 2012).

Within the ecological sciences resilience has two separate meanings that are often confused for each other, both having to do with how systems change through time and respond to those changes. The first refers to its older, more classical use where resilience describes the ability of a system to bounce back into a steady state point of equilibrium, thus maintaining the integrity of its design (Cretney, 2014). This definition was founded on a set of assumptions known as the "equilibrium paradigm" that drew upon the idea that stable, single-point equilibria are common in ecosystems. This particular definition of resilience is closely linked to its use in engineering practice, where its use may be appropriate for applications where there is a desired single stable state, such as bridge building. The equilibrium paradigm, along with "classic resilience", has now been largely discounted on account of the development of new ecological theoretical formulations.

The second definition of ecological resilience is based on a more inclusive, nonequilibrium paradigm. This view emphasizes the "dynamism, historical contingency and multiple pathways of change within an ecosystem" where external influences can have major structuring effects. In contrast to the equilibrium view, return to a fixed stable point of level within nonequilibrium resilience is not possible, as such states are not believed to exist within ecological systems. Stated simply, nonequilibrium resilience could be defined as "the ability (of a system) to adjust to changing, locally unstable conditions" (Pickett et al., 2014).

More recently, the concept of socio-ecological system resilience has been developed. Socio-ecological resilience is defined as the ability of groups or communities to cope with external stresses and disturbances as a result of social, political, and environmental change. In socio-ecological resilience, social and ecological systems are considered linked and interdependent of one another, having the potential to work through these linkages to either mutual benefit or detriment (Cretney, 2014). While not developed as a tool for studying cities exclusively, socio-ecological resilience models and concepts nonetheless translate well to urban sustainability studies.

There are two separate and defining elements to resilience theory that can be used to assess the resilience of any given system: adaptive capacity and transformation. These concepts have emerged "within socio-ecological resilience frameworks as elements of how complex systems behave and respond to challenges" and are "the patterns and processes of behavior that engage change to maintain a system within the parameters of critical thresholds" (Cretney, 2014). These processes include having the ability to learn and store information from past disturbances and the ability to prepare for and adapt to future uncertainty.

According to resilience theorist Carl Folke, "Adaptability captures the capacity of a socio-ecological system to learn, combine experience and knowledge, adjust

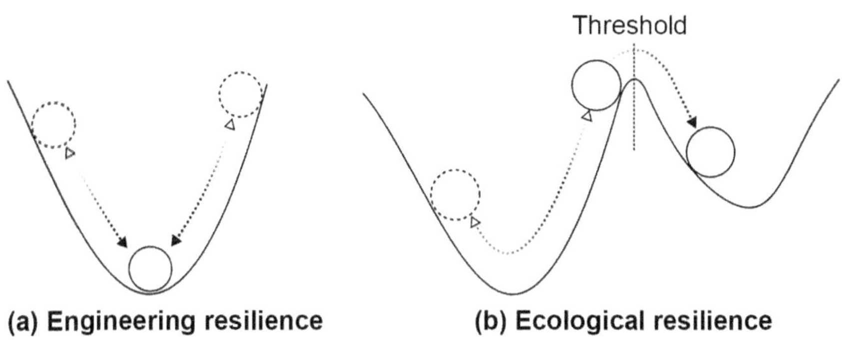

Figure 1.4 Ecological versus Engineering Resilience (Hoang, 2017)

its responses to changing external drivers and internal processes, and continue developing within the current stability domain or basin of attraction" (Folke, 2010). Thus, adaptability refers to the ability of people to determine the future course and development that a socio-ecological system takes, capable of avoiding disaster and remaining flexible, or in other words, resilient.

Transformation is then defined as the preparedness, option, and capacity to change when a current system becomes undesirable. Walker et al. further describe it as "the capacity to create untried beginnings from which to evolve a new way of living when existing ecological, economic, or social structures become untenable" (Walker B. a., 2012). This capacity for change is what distinguished resilience from general community capacities.

Resilience and urban design

In recent years, there has been an interest among urban designers and planners in expanding their understandings of urban processes to include natural phenomena and in doing so link them to broader environmental theory. Drawing from the insights of the field of urban ecology, planners now recognize the massive flows of energetic and material resources that move into and across urban ecosystem boundaries and the complex mosaics within them and the role that social and biophysical disturbances play in urban structure, change, and process.

Resilience theory, particularly as a metaphor, can be used as an effective tool to further bridge the fields of urban design and ecology. Multidisciplinary teams of ecologists, social scientists, and designers have begun dialog that can lead to design projects that will act as models of urban resilience. A number of scholars suggest using the adaptive cycle as general template for understanding how a city can adjust to internal and external shocks, as well as structural disruption (Pickett et al., 2014).

The idea of resilience as applied to urban environments has been further supported by a growing acknowledgment of the prevalence of urban transformations. While cities have been portrayed predominantly as unchanging, monumental architectural artifacts, in reality, they are subject to great changes. Many cities throughout history have suffered from massive depopulation resulting from forces ranging from economic collapse to environmental conditions, some recovering and some not. The Industrial Revolution also made profound shifts in the economies and populations of all cities it affected. In addition, the recent model of global development has resulted in many cities of the Global North becoming post-industrial, losing much of their employment, investment, and social cohesion in the process (Pickett et al., 2014). The adaptive cycle can be helpful in modeling these changes, particularly in the case of shrinking cities (Hartt, 2019).

Resilience criticism

Correspondingly to its widespread use and adoption in the fields of design and policy, the notion of resilient cities has also garnered substantial criticism. Much of this critique stems from the increasingly apolitical, eco-technocratic, and securitization-oriented nature of resilience discourse that tends to pay little more than lip service to matters of equity, justice, and power (Fitzgibbons, 2019); (Kaika, 2017). While some authors have gone so far as to reject the term outright (Ernstson H., 2014), others believe it still may have radical potential yet requires specificity regarding the type of resilience being advocated, for whom, and to what ends (Meerow, 2019). Problematic still is the matter of whether resilience is a normative property – do we want all things to be resilient? For instance, the economic system dependent upon the degradation of the planet and the exploitation of its peoples has been frustratingly resilient (Thorén & Olsson, 2018). What follows is an examination of key critical points raised in regard to resilience and its usage.

In *Resilience and Its Discontents*, Brendan Gleeson expresses his concern about applying the metaphor of resilience to cities. Unlike sustainability, which he considers to be a normative concept, Gleeson fears that the ambiguous nature of resilience may lead to it being used to promote authoritarian, regressive, neoliberal economic policies. Rather than being a tool to facilitate the development of sustainable cities, Gleeson fears that "resilience" has largely displaced "sustainability" as a favored term by both scientists and policymakers:

> Scientific displacement of sustainability has been accompanied by rising institutional and popular attachment to the resilience construct. In plain speak, it signifies the imperatives of premonition and pre-emptive response to threats and perturbations. In the language of contemporary policy interest it evokes the ideal of robustness in the face of shocks and crises thrown up by an increasingly disturbed natural order, and by complex disruptions in human systems such as the global economy.
>
> (Gleeson, 2013)

Gleeson sees the new enlistment of resilience thinking into urban planning policy as recasting the environment as "not so much in terms of benefits to be sustained for human benefit, but in terms of threats against which human well-being should be safeguarded". This, in his opinion, then sets the stage for a marriage of urban policy and security policy approaches.

Authors Mike Hodson and Simon Marvin echo Gleeson's concerns. They are troubled over the trend of "urban ecological securitization", a mindset that could lead to:

> transcending constraint through re-bundled ecologies taking the form of integrated responses to infrastructure that [move] across multiple infrastructure networks – energy, food, water, waste, etc. – and are rebundled together at particular scales in the design of new buildings, neighborhoods, towns, blocks, and cities.
>
> (Hodson, 2009)

These projects, the authors claim, focus on new developments that are located within or adjacent to existing cities, such as "eco-houses" or "eco-neighborhoods" that are much more concerned with integration at the scale of the development than with the wider transformation of the existing city or its incumbent infrastructure networks.

This characterization of urban ecological securitization meshes with concerns that Gleeson raises about the uncertain and ambiguous meanings of resilience, even between different academic disciplines. This lack of specificity, he states, can be overridden by the ideas' power and appeal to security thought and politics. This is concerning to the social sciences, whose "central precepts are anti-naturalism, and an expansive notion of human flourishing". To remedy this, Gleeson calls for greater communication between various disciplines and policymakers to understand the nuances of the term. "Unpacking this also requires improved understanding of the urban politics of resilience, the issues that become mobilized as crises or shocks, and the forms of knowledge and social interests that constitute such responses". His pleas for increased communication between natural and social sciences are similar to those being expressed by ecologists such as Pickett, who share similar concerns about the misapplication of resilience.

Any eco-scientific concept risks being seriously misappropriated when it is applied to social dimensions of life. Social Darwinism, the outstanding example, serves as a reminder of this danger. Similar caution needs to be taken with concepts such as resilience and sustainability.

Some scholars see resilience as being incommensurable with the social sciences. On account of the malleability of the concept, they've expressed concern that resilience could be used to scientifically justify policies and practices that could result in the depoliticization of social change (Olsson, 2014). For instance, poverty could be framed as being a stochastic condition rather than the product of political and structural forces. Furthermore, sociologists are wary of the principle of self-organization, a tenet of resilience thinking, being applied to social systems where it may obscure matters of agency, power, and conflict. They reject the notion of resilience as a grand unifying theory that adequately explains both natural and social processes, preferring a pluralistic approach where diversity is favored over grand theories that may suppress alternative views from historically marginalized sections of society.

Potential for grassroots resilience

While there are substantial criticisms of the concept of resilience and genuine concern over the ways it is being used to promote an agenda of neoliberalization and post-political ecological modernization, I believe the idea still has powerful potential to be used as a tool for building regenerative and equitable urban ecologies. The adaptive cycle, in particular, can be an effective framing device for challenging linear growth-fetishism and for embracing equitable urban degrowth models (Hamilton, 2004); (Hollander, 2011), seeing cycles of expansion, release, contraction, and regrowth not as inherently "good" or "bad" but rather as regular phases along a continuum of change. Furthermore, adaptability and transformability are desirable attributes of both people and places, particularly when viewed through an analysis of systemic oppression that recognizes how certain groups hold greater power and privilege than others. In an era of climatic shocks, food insecurity, pandemics, and increasing inequality, it is a noble goal to cultivate the resilience of environmental justice communities.

Brendan Gleeson moreover comments regarding the positive attributes of resilience:

> It can be summarized as the ability to disrupt the static, equilibrium-focused accounts of the human order, such as neo-classical economics, that have so demonstrably failed to account for natural ontology, human evolution, and, in recent history, the ruinous "progress" of neoliberalism. . . . Resilience also insists that the future is accounted and prepared for, not discounted through econocratic reason or trivialised through technocratic hubris. Its dynamic, adaptive view of history sits well with a progressive view that humanity is engaged in a massive project of co-evolution with nature, for which we bear responsibility as contributing authors of planetary fate.
>
> (Gleeson, 2013)

Recognizing its progressive potential, many alternative movements including low-impact urban developments, transition towns, and relocalization movements have incorporated the language of resilience, stressing local and community control of infrastructure and equitable access to environmental resources (Hodson, 2009). These groups represent more community-centered, decentralized, and grassroots interpretations of resilience. Operating on the lowest level of organization in a city, they nonetheless form a loose and potentially large system within the city that can contribute significantly to the social resilience of cities, despite the complexity of urban systems. Often, these organizations take on two different forms: those instigated by an immediate cause, such as a disaster (community responses to hurricanes Katrina and Sandy) and those that are driven by long-term sustainability goals such as food production, insulation of residential buildings, local energy cooperatives, and civil participation (Transition Town movement) (Meijer, 2013).

To facilitate resilience, an important relationship can be made between grassroots organizations and city government. One role that urban planners can play in this regard is to design opportunities for citizen self-organization and to facilitate connections between community organizations and city government. This type of cross-scale interaction will help cities achieve resilience by embracing modularity, interoperability, and distributed localism. Grassroots initiatives can play critical roles in facilitating the transfer of resources and information between city governments and the communities they serve. The collective efforts of such groups working in a community context has the effect of building greater resilience in the city as a whole.

Concluding thoughts on resilience

Resilience is a potentially powerful concept, particularly in its application in the ecological sciences to understand dynamic processes of ecosystem change and transformation. Resilience science overall helps in conceptualizing an ecology that embraces nonlinearity and multi-equilibria states, helping to explain how ecosystems can shift from one stable state into another and back again. In some instances, there may be potential to apply this thinking to socio-ecological systems, although this should be done with caution so as not to essentialize social processes or to risk depoliticizing them through "naturalization". Resilience, however, on account of its multiple definitions in ecology, engineering, and psychology has become a terribly nebulous term, used by many but understood by few. It has given rise to a new class of professional – the "resiliency expert" who can be hired by city governments or corporations to help increase their respective agencies' resilience, often expressed through quantified scoring. The confusion generated around resilience has led some to believe that resilience represents a type of augmented sustainability or "sustainability plus one". The difference between sustainability and resilience has also been described to me by a city planner as the formed being focused on outcomes, the latter was concerned more with process. To me, this separation seems arbitrary and fabricated, with nothing in any of resilience's multiple origins that would give credence to this distinction. While I believe it is important for communities to incorporate some of the concepts of ecological resilience theory into their planning, it should not be done as a replacement for sustainability thinking. While sustainability certainly has been heavily co-opted by numerous corporate and neoliberal interests, I believe it still has merit as a concept making it worth preserving. Sustainability, at least in its original definition, is explicit in its considerations of social justice and equity. This analysis is not explicit and is largely absent from most resilience thinking. More recent attempts to amend resilience with ideas such as adaptive capacity and transformation being done to put a progressive retroactive spin on what began as a nonnormative apolitical concept. More often than not, today, resilience is being used to push an agenda of risk-centered securitization that is compatible with the goals of urban neoliberalization (Wilson, 2004). While I don't believe that resilience is devoid of value or should be given up on entirely, I certainly would not advocate using it as a replacement framework or unifying concept for the numerous sustainability and justice initiatives well underway. As with any of these, the analysis needs to be sharpened to be explicit about the essential interdependence of environmental sustainability and social justice goals. From there, it may be possible to realize a truly "just resilience" (Pellow, 2018) that would be an integral aspect of an Urban Ecosystem Justice framework.

Notes

1 The term "anthropogenic" was first used by English ecologist and botanist Arthur Tansley. He used it to describe "anthropogenic climaxes" as a state of arrested ecological succession brought on through human activity, based on his ecological studies in English countrysides. The concept challenged Frederic Clements's notion of climax ecosystems, which was largely influenced by his observations of relatively "wild" landscapes in the American West. From the beginning, then, the use of the term anthropogenic challenged the human–nature dualism that was common in popular discourse since the beginning of the Industrial Revolution. Recognizing that even pre-industrial humans could

significantly shape the form and functions of ecosystems led to questioning the idea of nature as pristine and untouched. Tansley was however concerned about the potential consequences of anthropogenic activity, fearing it would spread "at an increasing rate, all over the face of the globe" and that all the earth's systems would in time all come under the "human yoke". Tansley, Arthur George. (1932). *Practical plant ecology. A guide for beginners in field study of plant communities*. London: George Allen and Unwin, Ltd.; New York: Dodd, Mead and Co.

2 This is in reference to indigenous settlements in the Amazon rainforest and specifically about the anthropogenic "Terra preta" or "dark earth" soils found in the region. Most of the soils in the Amazon lack sufficient fertility to support sustainable agriculture. The Terra Preta soils, however, contain higher levels of nitrogen, phosphorus, potassium, calcium, and organic matter than the surrounding native soils. These "anthrosols", consisting largely of composts, pottery shards, and charcoal, were created as a by-product of human settlements. They are also now being studied extensively for their potential to sequester atmospheric carbon, with the hopes that they could be replicated elsewhere. Lima, Hedinaldo N., et al. (2002). Pedogenesis and pre-Colombian land use of "Terra Preta Anthrosols" ("Indian black earth") of Western Amazonia. *Geoderma* 110(1–2), 1–17.

3 Agroecology explicitly centers the voices and interests of oppressed farming communities.

Works cited

Alberti, M. e. (2003). Integrating humans into ecology: Opportunities and challenges for studying urban ecosystems. *BioScience*, 53(12), 1169–1179.

Altvater, E. e. (2016). *Anthropocene or capitalocene? Nature, history, and the crisis of capitalism*. Oakland, CA: Pm Press.

Anderson, M. K. (2013). *Tending the wild: Native American knowledge and the management of California's natural resources*. Berkeley, CA: University of California Press.

Angus, I. (2015). Hijacking the Anthropocene. *Climate and Capitalism, 19*.

Asafu-Adjaye, J. e. (2015). *An ecomodernist manifesto*. Breakthrough Institute. Retrieved April 4, 2016, from www.ecomodernism.org

Atkins, P. J. (2012). *Animal cities: Beastly urban histories*. Farnham: Ashgate Publishing, Ltd.

Balée, W. a. (2006). *Time and complexity in historical ecology: Studies in the neotropical lowlands*. New York: Columbia University Press.

Barthel, S. a. (2013). Urban gardens, agriculture, and water management: Sources of resilience for long-term food security in cities. *Ecological Economics*, 86, 224–234.

Baskin, J. (2015). Paradigm dressed as epoch: The ideology of the Anthropocene. *Environmental Values*, *24*(1), 9–29.

Berkes, F. (2017). *Sacred ecology*. London: Routledge.

Boone, C. G. (2012). *Urbanization and sustainability: Linking urban ecology, environmental justice and global environmental change* (Vol. 3). New York: Springer Science & Business Media.

Botkin, D. B. (1992). *Discordant harmonies: A new ecology for the twenty-first century*. Oxford: Oxford University Press.

Brenner, N. P. (2009). Cities for people, not for profit. *City*, *13*(2), 176–184.

Broto, V. C. (2012). Interdisciplinary perspectives on urban metabolism. *Journal of Industrial Ecology*, *16*(6), 851–861.

Cadenasso, M. e. (2015). *The Baltimore School of Urban Ecology: Space, scale, and time for the study of cities*. New Haven, CT: Yale University Press.

Capra, F. a. (2014). *The systems view of life: A unifying vision*. Cambridge: Cambridge University Press.

Chelleri, L. e. (2016). Are people responsive to a more sustainable, decentralized, and user-driven management of urban metabolism? *Sustainability*, *8*(3), 275.

Childers, D. L. (2014). Advancing urban sustainability theory and action: Challenges and opportunities. *Landscape and Urban Planning*, *125*, 320–328.

Colding, J. S. (2019). Wicked problems of smart cities. *Smart Cities*, *2*(4), 512–521.

Cole, R. J. (2012). Transitioning from green to regenerative design. *Building Research & Information*, *40*(1), 39–53.

Collier, M. J. (2014). Novel ecosystems and the emergence of cultural ecosystem services. *Ecosystem Services*, 9, 166–169.

Collier, M. J. (2015). Novel ecosystems and social-ecological resilience. *Landscape Ecology*, *30*(8), 1363–1369.

Cretney, R. (2014). Resilience for whom? Emerging critical geographies of socio-ecological resilience. *Geography Compass*, 8(9), 627–640.

Crist, E. (2016). The reaches of freedom: A response to an ecomodernist manifesto. *Environmental Humanities*, *7*(1), 245–254.

Cronon, W. (1996). The trouble with wilderness: Or, getting back to the wrong nature. *Environmental History*, *1*(1), 7–28.

Crumley, C. L. (2018). *Issues and concepts in historical ecology: The past and future of landscapes and regions*. Cambridge: Cambridge University Press.

Davis, J. e. (2019). Anthropocene, capitalocene, . . . plantationocene? A manifesto for ecological justice in an age of global crises. *Geography Compass*, *13*(5), e12438.

Davis, M. (2007). *Planet of slums*. New York: Verso.

Decker, E. H. (2000). Energy and material flow through the urban ecosystem. *Annual Review of Energy and the Environment*, 25, 685–740 (Volume publication date November 2000), https://doi.org/10.1146/annurev.energy.25.1.685.

Del Tredici, P. (2010). *Wild urban plants of the Northeast*. Ithaca, NY: Cornell University Press.

Descola, P. a. (1996). *Nature and society: Anthropological perspectives*. London: Taylor & Francis.

DHS, U. S. (2020). *Resilience*. Retrieved from www.dhs.gov/topic/resilience

Dooling, S. (2009). Ecological gentrification: A research agenda exploring justice in the city. *International Journal of Urban and Regional Research*, *33*(3), 621–639.

Dull, R. A. (2010). The Columbian encounter and the Little Ice Age: Abrupt land use change, fire, and greenhouse forcing. *Annals of the Association of American Geographers*, *100*(4), 755–771.

Eisenstein, C. (2018). *Climate: A new story*. Berkeley, CA: North Atlantic Books.

Ellis, E. C. (2010). Anthropogenic transformation of the biomes, 1700 to 2000. *Global Ecology and Biogeography*, *19*(5), 589–606.

Engels, F. (1959). *The dialectics of nature*. Moscow: Progress.

Ernstson, H. (2014). Stop calling me resilient. *Comment on Tom Slater's blog post "The Resilience of Neoliberal Urbanism"*.

Ernstson, H. a. (2018). *Urban political ecology in the Anthropo-obscene: Interruptions and possibilities*. London: Routledge.

Fitzgibbons, J. a. (2019). Just urban futures? Exploring equity in "100 Resilient Cities". *World Development*, *122*, 648–659.

Folke, C. e. (2010). Resilience thinking: Integrating resilience, adaptability and transformability. *Ecology and Society*, *15*(4).

Frawley, J. a. (2014). *Rethinking invasion ecologies from the environmental humanities*. London: Routledge.

Girardet, H. (1992). *The Gaia Atlas of cities: New directions for urban living*. New York: UN-Habitat.

Girardet, H. (2008). *Cities people planet: Urban development and climate change*. New York: John Wiley & Sons Incorporated.

Gleeson, B. (2013). *Resilience and its discontents*. Research Paper No. 1. Melbourne: University of Melbourne, Melbourne Sustainable Society Institute. Retrieved September 20, 2015, from www.sustainable.unimelb.edu.au/sites/default/files/docs/MSSI-ResearchPaper-01_Gleeson_Resilience_2013.pdf

Gliessman, S. R. (1998). *Agroecology: Ecological processes in sustainable agriculture*. Boca Raton, FL: CRC Press.

Golubiewski, N. (2012). Is there a metabolism of an urban ecosystem? An ecological critique. *Ambio*, *41*(7), 751–764.

Grove, J. M. (2015). *The Baltimore School of Urban Ecology: Space, scale, and time for the study of cities*. New Haven, CT: Yale University Press.

Hamilton, C. (2004). *Growth fetish*. London: Pluto Press.

Hamilton, C. (2013). *Earth masters: Playing God with the climate*. Sydney: Allen and Unwin.

Haraway, D. (2015). Anthropocene, capitalocene, plantationocene, chthulucene: Making kin. *Environmental Humanities*, 6(1), 159–165.

Hartt, M. A. (2019). Resilient shrinking cities. In *The Routledge handbook of urban resilience*. New York: Routledge.

Harvey, D. (2008). The right to the city. *The City Reader*, 6, 23–40.

Heynen, N. M. (2006). The nature of cities: Urban political ecology and the politics of urban metabolism. In *In the nature of cities: urban political ecology and the politics of urban metabolism*. New York: Routledge.

Hoang, T. C. (2017). A systematic approach to the dilemma between flood vulnerability and resilience-review and concepts. *Vietnam Journal of Science and Technology*, *55*(5), 620. Creative Commons.

Hobbs, R. J. (2013). Defining novel ecosystems. In *Novel ecosystems: Intervening in the new ecological world order*. Oxford: John Wiley and Sons.

Hobbs, R. J. (2014). Novel ecosystems: Concept or inconvenient reality? A response to Murcia et al. *Trends Ecology Evolution*, *29*, 645–646.

Hodson, M. a. (2009). Urban ecological security: A new urban paradigm? *International Journal of Urban and Regional Research*, *33*(1), 193–215.

Hollander, J. B. (2011). The bounds of smart decline: A foundational theory for planning shrinking cities. *Housing Policy Debate*, *21*(3), 349–367.

Houston, D. (2013). Crisis is where we live: Environmental justice for the Anthropocene. *Globalizations*, *10*(3), 439–450.

Hughes, S. a. (2020). Just urban transitions: Toward a research agenda. *Wiley Interdisciplinary Reviews: Climate Change*, *11*(3).

Jenkins, J. (2019). *The humanure handbook*. Grove City, PA: Joseph Jenkins, Inc.

Kaika, M. (2017). "Don't call me resilient again!": The New Urban Agenda as immunology . . . or . . . what happens when communities refuse to be vaccinated with "smart cities" and indicators. *Environment and Urbanization*, *29*(1), 89–102.

Koch, A. e. (2019). Earth system impacts of the European arrival and Great Dying in the Americas after 1492. *Quaternary Science Reviews*, 13–36.

Kueffer, C. (2015). *Ecological novelty: Towards an interdisciplinary understanding of ecological change in the anthropocene*. Dordrecht, the Netherlands: Springer.

Kull, C. A. (2018). Critical invasion science: Weeds, pests, and aliens. In *The Palgrave handbook of critical physical geography* (pp. 249–272). Cham: Palgrave Macmillan.

Latour, B. (2013). "Facing Gaia": Six lectures on the political theology of nature. In *Gifford lectures on natural religion*, 18–28. Cambridge: Polity Press.
Levin, S. A. (1999). *Fragile dominion: Complexity and the commons*. Reading, MA: Perseus.
Lorimer, J. (2012). Multinatural geographies for the Anthropocene. *Progress in Human Geography*, *36*(5), 593–612.
Macintyre, B. (2010). The fault line in Haiti runs straight to France. *The Sunday Times*, 21.
Mann, C. C. (2005). *1491: New revelations of the Americas before Columbus*. New York: Alfred A. Knopf Incorporated.
Marris, E. (2013). Perspective: Coming of age in a trash forest. In *Novel ecosystems: Intervening in the new ecological world order*. Hoboken, NJ: John Wiley and Sons.
Martellozzo, F. e. (2014). Urban agriculture: A global analysis of the space constraint to meet urban vegetable demand. *Environmental Research Letters*, 9(6), 064025.
McClintock, N. (2010). Why farm the city? Theorizing urban agriculture through a lens of metabolic rift. *Cambridge Journal of Regions Economy and Society*, *3*(2), 191–207.
McNeill, J. R. (2016). *The great acceleration: An environmental history of the Anthropocene since 1945*. Cambridge, MA: Harvard University Press.
McPhearson, T. e. (2016). Advancing urban ecology toward a science of cities. *BioScience*, 6(3), 198–212.
McShane, C. a. (1997). The centrality of the horse in the nineteenth-century American city. In *The making of urban America* (pp. 105–130). Wilmington: Scholarly Resources.
Meerow, S. a. (2019). Urban resilience for whom, what, when, where, and why? *Urban Geography*, 40(3), 309–329.
Meijer, S. A. (2013). From community resilience towards urban resilience: Exploring the grassroot initiatives' role in cities. In *Living green scientific conference*. Delft, The Netherlands: Delft University of Technology.
Melosi, M. V. (2003). The historical dimension of urban ecology: Frameworks and concepts. In *Understanding urban ecosystems* (pp. 187–200). New York: Springer.
Mollison, B. (1988). *Permaculture: A designer's manual* (p. 576). Tyalgum, NSW: Tagari Publications.
Molotch, H. (1976). The city as a growth machine: Toward a political economy of place. *American Journal of Sociology*, *82*(2), 309–332.
Monstadt, J. (2009). Conceptualizing the political ecology of urban infrastructures: Insights from technology and urban studies. *Environment and Planning*, *41*(8), 1924.
Murcia, C. e. (2014). A critique of the "novel ecosystem" concept. *Trends in Ecology & Evolution*, *29*(10), 548–553.
Neyrat, F. (2018). *The unconstructable earth: An ecology of separation*. New York: Fordham University Press.
Odum, E. P. (1989). *Ecology and our endangered life-support systems*. Sunderland: Sinauer Associates.
Olsson, L. e. (2014). Why resilience is unappealing to social science: Theoretical and empirical investigations of the scientific use of resilience. *Science Advances*, *1*(4), e1400217.
Pellow, D. N. (2018). Afterword: From more than just sustainability to a more just resilience. In *Sustainability: Approaches to environmental justice and social power* (pp. 271–278). New York: New York University Press.
Perrow, C. (2011). *Normal accidents: Living with high risk technologies*. Princeton, NJ: Princeton University Press.
Pickett, S. T., McGrath, B., Cadenasso, M. L., & Felson, A. J. (2014). Ecological resilience and resilient cities. *Building Research & Information*, *42*(2), 143–157.

Pickett, S. T. (2016). Evolution and future of urban ecological science: Ecology in, of, and for the city. *Ecosystem Health and Sustainability*, *2*(7).
Plieninger, T. a. (2012). *Resilience and the cultural landscape: Understanding and managing change in human-shaped environments*. Cambridge: Cambridge University Press.
Redlinger, T. e. (2001). Survival of fecal coliforms in dry-composting toilets. *Applied and Environmental Microbiology*, *67*(9), 4036–4040.
Robbins, P. (2014). No going back: The political ethics of ecological novelty. In *Traditional wisdom and modern knowledge for the earth's future* (Vol. 1, pp. 103–118). Japan: Springer.
Ruddiman, W. F. (2010). *Plows, plagues, and petroleum: How humans took control of climate*. Princeton, NJ: Princeton University Press.
Ruser, A. a. (2016). Technology can save us, cant it? The emergence of the "technofix" narrative in climate politics. In *Proceedings of the International Conference Technology+ Society= Future*.
Scoones, I. (1999). New ecology and the social sciences: What prospects for a fruitful engagement? *Annual Review of Anthropology*, *28*(1), 479–507.
Sehgal, P. (2015). The profound emptiness of resilience. *The New York Times Magazine*.
Shellenberger, M. a. (2015). *An ecomodernist manifesto: From the death of environmentalism to the birth of ecomodernism*. The Breakthrough Institute.
SOIL. (2015). *What we do*. Retrieved January 11, 2015, from www.oursoil.org/
Stoetzer, B. (2018). Ruderal ecologies: Rethinking nature, migration, and the urban landscape in Berlin. *Cultural Anthropology*, *33*(2), 295–323.
Swyngedouw, E. (2006). Circulations and metabolisms: (Hybrid) natures and (cyborg) cities. *Science as Culture*, *15*(2), 105–121.
Thorén, H., & Olsson, L. (2018). Is resilience a normative concept? *Resilience*, 6(2), 112–128.
Tsing, A. L. (2014). Blasted landscapes (and the gentle arts of mushroom picking). In *The multispecies salon* (pp. 87–109). Durham, NC: Duke University Press.
Walker, B. a. (2012). *Resilience practice: Building capacity to absorb disturbance and maintain function*. Washington, DC: Island Press.
Walker, L. R. (2012). *The biology of disturbed habitats*. Oxford: Oxford University Press.
Weil, T. a. (2020). IT risk and resilience: Cybersecurity response to COVID-19. *IT Professional*, *22*(3), 4–10.
Wilson, D. (2004). Toward a contingent urban neoliberalism. *Urban Geography*, 5(8), 771–783.
Worster, D. (1988). *The ends of the earth: Perspectives on modern environmental history*. Cambridge: Cambridge University Press.
Worthy, K. (2013). *Invisible nature: Healing the destructive divide between people and the environment*. Amherst, NY: Prometheus Books.
York, R. a. (2003). Key challenges to ecological modernization theory: Institutional efficacy, case study evidence, units of analysis, and the pace of eco-efficiency. *Organization & Environment*, *16*(3), 273–288.

2 Exclusion of social sustainability and emergence of enviro-technocratic governance

"Sustainability", as it was originally defined, attempts to situate itself at the nexus of the environmental, economic, and social aspects of the world (Hess, 2009). Commonly represented as a Venn diagram with three overlapping spheres, these basic models give the impression that the three elements of sustainability are of equal importance to one another and ought to be given equal consideration in sustainability management decisions. There is much to be critical of in a such a simple framework, as pointed out by scholars of "critical sustainability" (Clark, 2018). Preferable models used by ecological economists depict ecological, social, and economic spheres as progressively smaller and nested within one another. These more accurately represent the ultimate dependence of economic activity upon the continuance of larger social and environmental processes and discourage the tendency to "externalize" social and environmental expenses (Daly, 2011). More recent sustainability models go so far as to depict "a just and safe space for humanity" that demonstrate the interdependence of society and global ecology by juxtaposing social parameters with a planetary boundaries framework (Raworth, 2012). Pushing the envelope further still have been calls to reject the ambiguity of sustainability by "situating" and "contextualizing" it within the context of specific histories, places, and intersectionalities (Sze, 2018). Regardless of its shortcomings, the classic sustainability model has made the idea of the "triple bottom line" commonplace in settings such as corporate boardrooms and city planning offices where at least lip service is paid to the idea of being mindful of environment, society, and the economy.

In recent years, however, a trend has emerged in these circles where a noted de-emphasis has been put on the social sphere of sustainability while greater importance is placed on economic and technical considerations (Gould, 2016). The result of this has been the sidelining of inconvenient considerations of equity, justice, fairness, power, access, race, and class as they pertain to urban sustainability conversations. Operating then under a depoliticized eco-friendly smokescreen, the "feel good" language of sustainability and its associated images of parks, gardens, and green transport are conjured to create the illusion of community consensus. Growth-driven initiatives are then permitted to proceed while never redressing the deep environmental inequalities and injustices that have pervaded the history of urbanization or asking the fundamental question: "sustainability for whom?". This new style of enviro-technocratic governance closely matches what has been

DOI: 10.4324/9780367858711-3

similarly described as "post-politics", a system where genuine debate and political discussion is deferred in favor of top-down planning that frames sustainability issues as purely technical matters separate from issues of justice (Swyngedouw, 2016). In this way, political resistance is subverted while neoliberal interests are rewarded by attracting global capital investment.

The manifestations of the desocialized enviro-technocracy are multiple. Ranging from "green gentrification" (Gould, 2016) to "urban ecological securitization" (Hodson, 2009), they share the characteristics of providing premium environmental services to the affluent while marginalizing the poor into neglected and polluted environments. In the paragraphs that follow I will explore their development within these unfolding "dreamworlds of neoliberalism" (Davis, 2011).

Running parallel to the critical discourse around enviro-technocracy is that around the concept of "just cities" (Marcuse P. e., 2009) and "just urban transitions" (Hughes, 2020). While the idea of "social justice" is frequently invoked, particularly in urban contexts, like sustainability it is often vague and misapplied. Just cities and just urban transitions literature attempt to define and specify the forms and processes through which justice manifests, particularly as it applies to the environment, climate, food, and energy. This section of the text will include a discussion of key ideas and concepts that are central to framing the explicit recentering of concerns of justice and equity to sustainability and how they relate to a broader framework of urban ecosystem justice.

Environmental justice

In many ways, urban ecosystem justice can be thought of as an outgrowth of environmental justice that expands its area of focus to include the biophysical components of the urban ecosystem. Environmental justice's analysis of issues of access and equality in terms of the distribution of environmental harms and services creates a solid philosophical platform upon which urban ecosystem justice is based. A combination of the civil rights and environmental health movements, environmental justice is (at times, confusingly) at once a concept, movement, and discipline that examines the disproportionate impact that toxic industries and land uses have on low-income people and communities of color. Furthermore, environmental justice has done much to challenge traditional views of what constitutes "the environment", expanding environmental definitions to include non-pristine spaces such as industrialized and polluted inner-city environments along with their social ongoings. By doing so, the environmental justice movement has successfully made environmentalist tools previously used exclusively by middle class and white organizations accessible to many low-income people and communities of color. For the remainder of this section, I will explore the history and meanings of environmental justice, analyzing its ground-level applications and finally explaining its relation to the emerging notion of just sustainability.

Environmental justice began more than thirty years ago as a synthesis of the environmental movement and the social justice movement, arguably two of the most significant activist movements of the twentieth century (Frumkin, 2005).

The movement has multiple foundations that include the civil rights movement, academia, antitoxics, labor, indigenous rights, the academy, and the environmental movement. This organic emergence of environmental justice has made it difficult to pin an exact single beginning time and place on it (Cole, 2001). Environmental justice grew as a response to the tendency of polluting industries to many times either locate or be located in low-income communities of color, thus subjecting this population to a disproportionately high level of exposure to toxic burdens. As stated by Robert D. Bullard, the "father" of the environmental justice movement,

> [environmental justice demonstrates how] blacks, lower-income groups, and working-class persons are subjected to a disproportionately large amount of pollution and other environmental stressors in their neighborhoods as well as in their workplaces.
>
> (Bullard, 2008)

In many instances, communities facing sources of stress beyond their control and living within proximity to polluting industry, the impact is worsened by the role of economic and racial discrimination (Council, 2004).

Frumkin (Frumkin, 2005) lists three concepts as central to the study of the environmental justice movement. These are:

(1) The meaning of disproportionate impacts on populations that includes cumulative environmental hazards, vulnerability, inequalities in enforcement of regulations, and disparities in people's socioeconomic standings.
(2) The research and policy challenges involved in incorporating social justice concepts into an environmental law paradigm.
(3) The collaborative problem solving and community-based tools and tactics necessary to address the interwoven health, ecological, economic, and social issues facing underserved and overburdened populations.

While environmental justice primarily consists of a grassroots activist and local movement, it is also regarded distinctly as a policy principle and concept carried out by governmental agencies. Within this area of policy, considerable debate exists as to how it is implemented, its precise meaning, and the relationship it creates between scientific "experts" and the knowledge and experience of the public.

The EPA defines environmental justice in the following way:

> Environmental Justice is the fair treatment and meaningful involvement of all people regardless of race, color, national origin, or income with respect to the development, implementation, and enforcement of environmental laws, regulations, and policies. EPA has this goal for all communities and persons across this Nation. It will be achieved when everyone enjoys the same degree of protection from environmental and health hazards and equal access to the decision-making process to have a healthy environment in which to live, learn, and work.
>
> (USEPA, 2020)

This definition, while reasonably accurate, is nonetheless filled with terms that have ambiguous meanings that are highly dependent on the context is which they are used. Concepts such as "fairness", "justice", and "meaningful involvement" are all highly fluid concepts subject to broad interpretation (Harding, 1991). The idea of justice alone has multiple meanings. Modern conceptualizations include the distributive, procedural, cultural, and representative models, all representing varying components of an overall framework for justice (Allen, 2013).

In recent years, the field of environmental justice has expanded beyond solely collecting empirical data to demonstrate that low-income communities of color suffer disproportionately to toxic exposures (Holifield, 2009). The discipline now looks at a variety of socio-natural injustices that include "natural disasters" and transportation issues and has crossed disciplinary terrain with fields such as critical race theory and urban political ecology (Swyngedouw, 2009). Environmental justice scholars now regularly consider factors other than race, such as gender, ability, age, and sexuality when investigating the causes and impacts of social vulnerability.

The environmental justice movement has also begun a process of globalizations. Although it has its origins in cities in the United States, environmental justice movements now exist on multiple continents and examine the effects of the distribution of toxins over a broad area. For these reasons, certain scholars have begun to bring environmental justice insights into communication with the discipline of urban political ecology, a field that is concerned with the political and economic reworking of human and non-human relationships and the production of socioeconomic inequality (Swyngedouw, 2009).

The notion of environmental justice also has been criticized for failing to capture the complexity of the many political, ecological, biological, and medical aspects of the issue. A generalization of humanity occurs in the process of conceiving of environmental justice broadly, although injustices happen to specific groups of people in specific locations with their own unique conditions of climate, geography, history, and politics. These specific conditions can make it difficult to isolate exact causes of illness, a result also stemming from the problems created when applying general theories to ground-level situations (Fortun, 2011).

This complexity and uncertainty can manifest as a lack of trust between affected populations and scientific experts. Communities may not always be certain that professionals are sharing all the information they possess. While there are many literatures that discuss the relationship between experts and the public in the context of environmental justice (Allen, 2013); (Frickel, 2011), less of it prescribes methods for direct engagement of the public. One method of working amidst this high level of uncertainty and complexity is through the implementation of techniques that combine local insights or "contextual knowledge" with professional techniques. This environmental health framework is referred to as "street science" by Corburn (Corburn, 2005) and can serve to help community-generated information useful for improving environmental health decision-making. Recently, a field of inquiry called "critical environmental justice" has emerged that applies a more intersectional lens to environmental justice issues, incorporating enhanced analysis of gender and non-human dynamics in environmental justice work, as well as examining the EJ dimensions of Black Lives Matters and of Prisons (Pellow, 2017). This scholarship charts a path forward to a more contextualized and situated environmental justice.

Just sustainability

Conventional discourses on sustainability, sustainable development, and urban sustainability focus primarily on issues of environmental conservation, often originating from a top-down, centralized and specialized perspective. Adherents of this environmental sustainability paradigm are more frequently white and of upper- or middle-class socioeconomic status. The focus within this movement is on wilderness conservation, climate change prevention, biodiversity preservation, and other issues that largely frame the environment as something external and abstract. Concerns regarding equity and justice are less commonly expressed in this movement.

In contrast, the environmental justice movement comprises members representing more racially and economically diverse backgrounds and is focused more typically on resisting toxic industry and the unequal distribution of environmental harms. Far greater emphasis is put on issues of equity, collective action, and an overall sense of community. Despite the many successes of the environmental justice movement, it has been critiqued for being largely reactive, unable to take steps toward "developing consensual visions and taking ownership of the assets and resources necessary to bring such visions to fruition" (Agyeman J., 2005) on account of being strapped for resources and perpetually on the defensive.

These two movements represent two extremes on a continuum of responses to environmental issues that take issues of equity into consideration. Apart from a few short-lived collaborations, these movements remain separated, despite sharing many goals in common.

In response to the disparity, Julian Agyeman has coined the term "just sustainabilities" to describe a vision of sustainability that can be useful in bridging the gap between environmental justice and mainstream sustainability movements. This platform takes into account the "equity deficit" that pervades much of green and environmental thinking. According to Agyeman,

> Sustainability cannot be simply a "green", or "environmental" concern, important though environmental aspects of sustainability are. A truly sustainable society is one where wider questions of social needs and welfare, and economic opportunity are integrally related to environmental limits imposed by supporting ecosystems.
>
> (Agyeman J. R., 2002)

Agyeman sees a just sustainability platform as combining the best of these two movements. Just sustainability keeps issues of access and equity in the center yet is simultaneously reactive and proactive, "operating within an environmental justice framework but . . . also exploring the wider and emerging terrain of sustainable development and the development of sustainable communities" (Agyeman J., 2005).

While the just sustainability concept itself does not seem to have been widely adopted within circles of academics or activists, the core essence of its ideas and vision is becoming increasingly more adopted by communities looking to bridge the divide between environmental activism and social justice work. The central values it espouses are integral to an urban ecosystem justice platform, and it is my wish that urban ecosystem justice be thought of as an outgrowth of just sustainability thinking.

Green gentrification

Described as the convergence of "urban redevelopment, ecologically minded initiatives and environmental justice activism in an era of advanced capitalism" (Checker, 2011), green gentrification or environmental gentrification is the process where lower income people and communities of color are displaced from their homes and neighborhoods on account of increases in property values resulting from the creation of environmental amenities such as parks, community gardens, and environmentally sustainable housing primarily intended to benefit white and middle- to upper-class residents. Because of this displacement, lower-income individuals are forced to move into more affordable areas that experience higher burdens of polluting and toxic industries. Green gentrification can be explained as a result of failing to actualize the social equity component of sustainability, resulting in disconnected feedback between the natural systems and social systems of the city. Wealthier residents benefiting from environmental privilege by living in close proximity to environmental "goods" are less aware of polluting industries and environmental toxicity and are less motivated to act upon them. Poorer individuals, in contrast, suffer disproportionately more from exposure to toxins and ecological degradation, affecting their health and capacity to bring about change (Gould, 2016).

It has been pointed out that many times environmental gentrification comes about paradoxically as a consequence of successes of the environmental justice movement. By successfully resisting polluting industries and by achieving ecological remediation, neighborhoods are made more attractive to affluent residents. On a discursive level, urban planners will often appropriate the language of sustainability used by environmental justice activists to create a consensus around seemingly environmentally and socially conscious developments, while they are in truth "subordinat(ing) equity to profit minded development" (Checker, 2011).

As seen, well-meaning endeavors intended to improve the health and resilience of urban ecosystems can result in the displacement and disenfranchisement of the very communities they were intended to benefit. It is critical that concerns regarding social equity and justice remain explicit and in the forefront of sustainability conversations. Particularly, "Process equity" (Agyeman J., 2005) must be carried out when environmental improvements are made in a community. Process equity ensures that affected individuals participate in the design and implementation of a project from start to finish and that their ideas and concerns are given genuine consideration. Furthermore, in keeping with a holistic view of sustainability as incorporating both environmental and social considerations, urban sustainability activists concerned about displacement and gentrification should be simultaneously advocating for public policy initiatives favoring affordable housing along with urban greening initiatives. Such a dual power strategy is in keeping with the ethic of the "hatchet and the seed", whereby activists are at once both reactive and proactive (Robbins, 2011).

Yet another approach to addressing environmental gentrification is the idea of making cities "just green enough" (JGE) (Curran, 2012); (Wolch, 2014), characterized as a development approach that seeks to clean up toxic sites while simultaneously preserving industries and blue-collar work. Such an approach specifically eschews "parks, cafes, riverwalks", and other such high-end projects that could spur further real estate development and result in economic displacement. Another component of the JGE approach is to favor distributed and scattered small-scale greening endeavors as opposed to grand-scale and geographically concentrated greening projects (such as the reuse of mines or oil refineries). By doing this, the hope is that mega projects would not become focal points for further property development. JGE advocates admit that such practices require extraordinary cooperation between a wide array of community stakeholders and involve resisting the pressures of business and real estate. They would also need to be enacted in concert with affordable housing policies and rent control measures if they're to be effective. Gould and Lewis (Gould, 2016) are skeptical of the effectiveness of a JGE strategy in global cities, wary that the "green-gentrification tipping point" in "hot market" cities such as NYC is much more easily crossed than in "cold market" cities such as Detroit and Gary, Indiana. According to them, greening endeavors such as community gardens in the latter would result in a "better long-term environmental justice trajectory".

A persuasive argument has been made that stable, equitable, and inclusive neighborhoods are most resistant to gentrification and that strategies to improve communities and remove properties from market speculation should be taken regardless of the economic conditions of the neighborhood (Axel-Lute, 2019). The establishment of land trusts, community-owned gardens, and cooperatively managed green spaces can make property unavailable to speculators and provide stability and control to long-term residents. Such strategies are examples of commons-based systems advocated for by the "Right to the City" framework.

The just city and the right to the city

Before going deeper into the urban ecosystem justice concept, it is worthwhile to delve more into definitions and understanding of justice, particularly regarding the literature on the idea of the "just city". Originating with "The Republic" by Socrates, beliefs about urban justice have come to be centered on distributions of political and economic power as well as the effect of spatial inequities among a city's population as expressed through racial and class distinctions. Prominent among urban justice scholars today is David Harvey, whose Marxist inspired 1973 book "Social Justice and the City" serves as a touchstone (Harvey, 2010). Harvey's cohort charts the ascent of neoliberal capitalism in the 1970s which triggered a shift in the urban condition that was characterized as promoting privatization and deregulation in order to increase economic efficiencies. These policies had the effect of compounding problems of poverty by concentrating power into the hands of an elite minority at the expense of the majority.

A particular literature pertains to the role that the just cities concept can play in urban planning (Fainstein, 2010) and how the state interplays with citizens, producing injustices for marginalized peoples. Despite the fact that urban planning was originally created to address material inequities through spatial interventions, it in many cases has compounded these disparities on account of its institutionally ingrained inclination to support growth over the well-being of the public (Miraftab, 2004). Building off of Fainstein's ideas is the nascent notion of the climate-just city, which seeks to incorporate concerns of social justice (people), environmental justice (places), and ecological justice (non-humans) into a progressive planning agenda (Steele, 2012). The climate-just city platform draws from the broader idea of climate justice, a movement that professes that the responsibility for climate change is not equally distributed and that certain vulnerable groups will be affected more severely by its consequences (Barnett, 2006). It displays interesting potential synergy with urban ecosystem justice in that it asks how notions of justice can be extended to take into consideration the rights of non-human life and of ecosystems themselves, although UEJ would more greatly emphasize the importance of human–ecosystem relationships being necessarily interdependent and reciprocal.

While urban ecosystem justice and the idea of the climate-just city overlap considerably, the latter is primarily intended to be utilized for city planning and other top-down approaches. While just-planning is certainly a component of urban ecosystem justice, a greater emphasis is put on the agency of the average urban resident and nonspecialist to build ecological relationships and to affect change. In line with this sentiment, the idea of "the right to the city" as developed by Henri Lefebvre has particular relevance to the urban ecosystem justice framework (Lefebvre, 1972). Originating as a demand for people's access and control over everyday urban life, the idea of "the right to the city" was coined in France in the 1960s in response to widespread student protests in the streets of Paris. The term was later revived by geographer David Harvey in 2008 and was seized upon in subsequent years as a protest cry by movements ranging from Occupy Wall Street to the Arab Spring (Harvey, 2008).

Harvey explains the idea in the following way:

> The right to the city is far more than the individual liberty to access urban resources: it is a right to change ourselves by changing the city. It is, moreover, a common rather than an individual right since this transformation inevitably depends upon the exercise of a collective power to reshape the processes of urbanization. The freedom to make and remake our cities and ourselves is, I want to argue, one of the most precious yet most neglected of our human rights.
>
> (Harvey, 2008)

Marcuse takes the idea further, asking critical questions such as "who's right?" and "what city?" that are unanswered by the slogan alone (Marcuse P., 2012). Asking this is of particular importance as a wide range of groups representing different populations and political agendas have adopted the mantra as a rallying cry. Prominent among them is the organization "Right to the City", which focuses on

> building a national movement for racial justice, urban justice, human rights, and democracy . . . born out of an idea of a new kind of urban politics that asserts that everyone, particularly the disenfranchised, not only has a right to the city, but as inhabitants, have a right to shape it, design it, and operationalize an urban human rights agenda.
>
> (RTC, 2017)

The Right to the City movement espouses a view that regards urban space as common property, believing housing and public space should be free from speculation and instead be used for community building and cultural and political space. This sentiment is well summarized by the popular slogan of the movement, "cities for people, not for profit", and takes a hard stance against gentrification and displacement. As environmental justice additionally comprises one of the main philosophical pillars of the movement, Right to the City and urban ecosystem justice share much in common. Urban ecosystem justice would extend this analysis of housing and public space to include systems such as air, water, soils, and biodiversity. The Right to the City manifesto also includes provisions on rural justice, which would indicate a willingness to build solidarity beyond urban boundaries with oppressed rural residents.

Shrinking cities

Informed by the logic of capitalism, conventional wisdom holds that a healthy city remains in a constant state of growth and that a shrinking city is an unhealthy one. Society's fetishization of growth has produced a "metropolitan bias", resulting in most urban studies focusing on larger, more economically influential "global" cities such as New York City, Chicago, San Francisco, and Los Angeles, with occasional discussion of a second tier that includes cities such as Boston, Seattle, Atlanta, Miami, and Phoenix. Almost never considered and frequently outright ignored is the "shrinking city", defined as "densely populated urban area with a minimum population of 10,000 residents that has faced population losses in large parts for more than two years and is undergoing economic transformations with some symptoms of a structural crisis" (Wiechman, 2007). These cities are plentiful – globally as many as one in six cities can be said to be shrinking (Herrmann et al., 2016). Over 450 cities in the world with populations above 100,000 have lost at least one-tenth of their population since the middle of the century, including 59 in the United States (Oswalt, 2006). While diminished manufacturing is the primary cause of shrinking city's population decline, other contributing factors are disasters, war, political collapse, aging populations, and suburbanization

(Blanco, 2009). As little attention has been paid to the phenomenon of shrinking cities, city planners have been left ill-equipped to deal with urban contraction.

Many of these shrinking cities in the United States are proverbial "rust belt" towns and include places such as Syracuse, Flint, Youngstown, Akron, Rochester, and Worcester. With their downtowns deserted and manufacturing industries long gone, these smaller cities are further characterized as being economically depressed, racially segregated, crime stricken, culturally bland, divided by interstates, and environmentally contaminated. In contrast to this purely negative depiction, *Small, Gritty, and Green* (Tumber, 2011), author Catherine Tumber makes the case that these smaller post-industrial and shrinking cities have great potential to be revitalized, citing the fact that smaller cities as a class have many attributes that make them more desirable for implementing strategies for low-carbon and sustainability transitions.

The authors of "Ecology for a Shrinking City" (Herrmann et al., 2016) concur with Tumber, pointing to the fact that smaller cities have land assets within their borders – open space and vacant lots that could be used for agriculture, composting, solar farms, biodiversity conservation, rainwater infiltration, climate mitigation, and other ecosystem services. As such, they seek to bring an *ecology for the city* approach to rust belt towns, working with local residents to determine how vacant space can be made into a community asset that also improves urban ecosystem health. Closeness and proximity of this green space to residential and industrial centers permits an interchange and interaction that would not be possible in larger cities where open space is at a premium or nonexistent altogether.

Shrinking cities are additionally ideal for urban ecosystem justice practice due to the fact that smaller urban to rural gradients exist in smaller city regions. Lacking the miles of suburban sprawl that has consumed rural and wild space in the periphery of larger cities, the relatively quick transition from urban to rural space again promotes greater interchange between these two environments. This has the combined effect of giving rural residents easier access to the cultural opportunities provided by cities, and likewise makes farms, fields, and forests more within reach of city dwellers. Where the capacity for interchange maximized, shrinking city's reduced gradient could also ameliorate the polarizing and hurtful town–country dichotomy that results in significant political misunderstandings. This, of course, is by no means guaranteed and would require very active work on the part of boundary spanning activists to illustrate how (particularly low-income) urban and rural residents have many shared concerns.

Yet another attribute of shrinking cities is the large amount of relatively intact industrial infrastructure that *has the potential* to be repurposed, at least in part, for the construction of solar panels, wind turbines, and other equipment necessary to bring about a renewable energy transition. Many of these cities still retain a population capable of acting as a workforce in these assembly plants. While many times entirely new factories are constructed to produce renewable energy equipment, the reuse of older infrastructure should be encouraged to whatever extent possible. Lastly, older industrial cities commonly used to locate along waterways to harness hydropower for running their operations. In some instances, this hydro infrastructure is still present and could be brought back on line as a renewable energy supply.

Ecological gentrification in the shrinking city

At first glance, it may appear that cities with overall negative population growth are at a reduced risk for being gentrified. Gould and Lewis make the claim that in shrinking cities

> [U]rban greening is unlikely to substantially increase the potential to appropriate environmental resources up the stratification scale . . . greening might stabilize property values, not hyper-inflate them. . . . Converting vacant land in Gary or Detroit into an urban farm . . . will increase or stabilize property values, but is not going to attract developers looking to a global market for luxury condominiums in urban centers. . . . Greening environmental-justice communities in cold markets is going to have a better long-term environmental-justice trajectory than doing so in global cities.
>
> (Gould, 2016)

While there is certainly truth in this observation, it does not mean that gentrification concerns can be dismissed outright. Gould and Lewis concede that urban greening could affect property values in the "hottest" neighborhoods in shrinking cities, which could result in lower income residents displaced on to the urban periphery, which in a shrinking city is increasingly underserved. The authors of "Ecology for a Shrinking City" acknowledge this potential, emphasizing strong community involvement grounded in and led by local residents as necessary to avoid a localized gentrification (Herrmann et al., 2016).

Challenges and responses to shrinking cities

The metaphor of "shrinkage" has been criticized (Nassauer, 2014) as being inaccurate as shrinking cities are not growing smaller in terms of geographic area but rather in population. As a consequence of them contracting, shrinking cities have left behind a large amount of "legacy infrastructure" – roadways, water mains, gas pipes, power lines, etc. – which must still be maintained to service areas that are no longer populated. Servicing this vast expanse of infrastructure becomes a financial burden for cities as their tax bases decrease. Rather than holding out hope that their peak population numbers would return and despite the reluctance on the part of city governments and planners to accept that their cities are shrinking, a handful of cities have embraced urban contraction as a planned pathway (Popper, 2010). Embracing degrowth entails concentrating urban services into downtown areas, removing highways, enhancing walkability, and extensively demolishing abandoned structures and housing stock. The degrowth mindset includes an understanding that perpetual growth is not equal to urban development and that slow growth, stability, and managed decline are all acceptable urban strategies

(Hollander, 2011). These multiple urban states can be effectively visualized using the adaptive cycle, a tool commonly employed in resilience science (Gunderson & Holling, 2002). Any of the points along the R to K continuum, including the "back loop", are equally as legitimate, provided appropriate policies and responses are employed (Pelling & Manuel-Navarrete, 2011). In this way, a nonequilibrium ecological paradigm that upends the notion of climax ecosystems can be used to challenge the doctrine of the city as a "growth machine" (Molotch, 1976).

Foremost among them is Youngstown, Ohio, which in the early 2000s adopted a policy of "smart decline" to manage its shrinkage. Its *Youngstown 2010 Plan* detailed the city's ambition to "right size" itself following years of continued population loss (Schatz, 2013). While initially praised by many as being forward thinking and "almost revolutionary" (Wiechmann, 2012), the implementation of the Youngstown Plan has now been criticized for falling short of many of its stated goals and for disproportionately benefiting the wealthier residents of the city (Rhodes, 2013). By pursuing exclusively market-oriented solutions to counter its shrinkage, Youngstown failed to address and therefore compounded the underlying social issues and inequities that are entangled with and contribute to population decline. For example, in an attempt to reduce the physical size of the city, Youngstown enacted a form of "neighborhood triage" where different areas of the city were determined to either be capable of being revitalized or not. Those that were deemed redeemable benefited somewhat in improved services, while those considered beyond salvation had services cut, were widely demolished, and permitted to continue to deteriorate. In this sense, conditions for the city's poorest residents worsened through degrowth policies. Furthermore, new businesses were incentivized to move to the city yet provided high-skill jobs for only those who were already upwardly mobile.

Many smart-decline policies seem to share much in common with what is described as "urban austerity", a condition of neoliberalism imposed upon cities characterized by privatization, disinvestment in the public sector, asset sell-offs, and the reduction of pensions and workplace rights (Peck, 2015). Most notably affected is the city of Detroit, which in 2013 was assigned an emergency manager by the republican Michigan government who was tasked with balancing the city's budget. While ultimately declaring bankruptcy for the city, imposed austerity measures included cutting off essential services such as electricity and water to many of the city's (poorest) residents. Great care must be taken in managing urban contractions that the benefits and sacrifices of downsizing are equitably distributed among all city residents.

In conclusion, shrinking cities present an interesting array of challenges and opportunities. Unburdened by the persistent downward economic pressures present in hyper-capitalistic "global" cities, there exists greater space and time to pursue the development of equitable and just urban socio-ecological relationships that might otherwise be impeded or made impossible in quickly growing cities. In the absence of tensions related to rapid increases in property values and the subsequent displacements created by them, communities may have greater success in building relationships of trust and mutual support that span racial and economic boundaries. While communities must certainly remain vigilant about the potential for ecological gentrification to occur (especially on a localized neighborhood level), the overall risk is apparently reduced – with overall lowered property values, there is a lessening of the sense that trade-offs between either green space or affordable housing must be made. The abundance and accessibility of vacant space creates expanded potential for urban agriculture and the providence of ecological services, particularly with the close integration of obsolete industrial infrastructure and the relative closeness of rural space. Reduced urban-to-rural gradients can facilitate enhanced interchange and understanding between urban and rural communities that may otherwise be polarized and turned against one another for the purposes of political gain.

Despite their potential, shrinking cities are commonly saddled with a collective cultural attitude of despair and hopelessness that is persistent, difficult to overcome, and may inhibit the implementation of innovative strategies. Enormous attention needs to be paid to patterns of injustice and inequity that have historical roots in many post-industrial cities and continue to manifest to this day. Great care must be taken to honor the communities that have remained in shrinking cities and have built networks of support over the years – the tendency to see shrinking cities as wastelands and therefore as blank slates for improvement carries with it ingrained attitudes of cultural superiority and colonialist mentalities of salvation. It is possible to manage the contraction of an urban area in a manner that is equitable, just, and environmentally sound – determining the best practices for doing so is a critical component of sustainable urban futures that run counter to visions of cities as pure neoliberal growth engines.

The urban commons

The notion of urban ecosystem justice also includes an analysis of the idea of "the commons" and specifically its extension into the realm of the city as "the urban commons". The commons are defined as being a pool of resources, such as forests, fields, or water bodies, jointly used and managed by a community of people accordingly to their own community defined rules (Siefkes, 2009). These rules are typically structured so as to be relatively egalitarian, incorporative, and fair (Gidwani, 2011) and prevent resources from becoming privatized or misused. In contrast to the idea of "the tragedy of the commons" by ecologist Garrett Hardin (Hardin, 2009) and supported by the work of Elinor Ostrom (Ostrom, 2015), commons are not automatically doomed to overuse and degradation but rather can be sustained for generations through the careful management of "the commoners" or those who participate in the use and governance of the commons. In this sense, it is important to distinguish between "open access" resources and "the commons", with the former having no regulations or oversight and therefore being vulnerable to degradation. Furthermore, the difference between the terms "public" and "the commons" must be explained as they are frequently used interchangeably.

"Public" is a legal term used by state authorities to make a distinction between it and that which is "private". The commons

> historically and etymologically, are that which lie at the frontiers, or within the interstices, of the territorial grid of law. They exist as a dynamic and collective resource – a variegated form of social wealth – governed by emergent custom and constantly negotiating, rebuffing, and evading the fixity of law.
>
> (Gidwani, 2011)

Despite their different meanings the terms are not mutually exclusive, and much of what is regarded as public, such as parks, could also considered to be part of the commons.

The destruction of the commons system is regarded as being a necessary prerequisite for the expansion of capitalism and neoliberal economic policies. The dispossession and deprivation of populations dependent on common resources initially entails an "enclosure" and privatization of common resources followed by their subsequent degradation. Displaced populations are then forced to migrate to urban environments where they become dependent on industrial wages for survival (Gidwani, 2011).

While most commons historically have been regarded as being rural artifacts, recently there has been a shift in perceptions to include urban spaces and customs in the subject category of "urban commons". Sharing a similar outlook to "right to the city" in posing a significant philosophical challenge to dominant neoliberal urbanist discourse, both it and the urban commons concept consider public goods and resources such as parks, schools, transportation, the breathable atmosphere, forests, waterways, streets, wetlands, gardens, plazas, etc. to belong collectively to the residents of the city. Additionally, notions such as public culture itself and urban waste streams could be regarded as belonging to the commons as well. In this regard, there is a mix of common resources that could be categorized as both ecological and civic. In many instances, these commons are rapidly diminishing as they are replaced by enclosed, privatized, and monitored public spaces such as malls and gated environments (Gidwani, 2011). Preservation of the urban commons is critical as they provide local access and control over essential resources to urban populations, particularly in the Global South where they provide a safety net for recent urban immigrants (Nagendra, 2014). Furthermore, commons provide training grounds for practicing the democratic management of resources and for providing spaces for the creation and preservation of communities themselves.

Another body of literature explores what it refers to as the "urban green commons" or UGC. These are described as "physical green spaces in urban settings of diverse ownership that depend on collective organization and management and to which individuals and interest groups participating in management hold a rich set of bundles of rights" (Colding J. e., 2013). These rights include the ability of participants to create their own governance institutions and choose whom to include in management systems. Some examples of UGCs include allotment gardens and community gardens. These are more closely parallel "traditional" commons such as rural fields and woodlots in that they do not "belong to everyone", resources may be "extracted" from them, and that rights of exclusion are a key component of participation. In this sense, they are clearly distinguished from other more "open access" forms of urban commons such as the atmosphere, which genuinely belongs to no one and is made use of by all.

The case is made that UGCs provide important environmental services in cities including management of cultural diversity, promoting social-ecological memory, cognitive resilience building, reducing economic vulnerability, and representing institutional redevelopment designs for shrinking cities (Colding J. a., 2013). Interestingly, affordability and accessibility are not necessary attributes of UGCs as defined by Colding et al., as they leave open the possibility for collectively managed gated communities to be considered as a form of UGC. For this reason, and by the author's own admission, there is a need to "elaborate on how (UGCs) can come to animate a wider debate in cities on who can claim knowledge concerning how, where, and by whom urban nature can be managed and protected" (Colding J. a., 2013).

In conclusion, it appears as if study and discussion of the phenomena of the commons is gaining popularity in academic and nonacademic circles. It warrants greater consideration as a governance model, particularly in the urban realm where it provides a citizen-centered alternative to neoliberalism other than state-sponsored socialism. I will reference the idea of the commons on several instances in this book, particularly as it pertains to areas of the urban ecosystem such as waters, soils, atmospheres, and waste streams.

Conclusion to part I

In this introduction I have presented my central thesis of urban ecosystem justice and have performed an examination of a number of concepts, both ecological and social in nature, that together create a tile mosaic of ideas, philosophies, and possibilities that collectively form the chaotic bricolage of urban ecosystem justice. A response to the interrelated challenges of urban ecological alienation and the decentralization of social sustainability, urban ecosystem justice seeks to promote greater equity, inclusion, justice, and access in the urban ecosystem by demonstrating the relevance of its biophysical dimensions of soil, water, air, waste, and biodiversity to historically underrepresented urban populations. Beginning by unpacking the history of nature–society dualism and reductionist scientific thinking, I explored the potential for reunification through anthropogenic ecosystems, the Anthropocene, resilience, urban ecology, urban metabolism, and novel ecosystems. Moving on then to an examination of more sociological concepts, I delved into the meanings and histories of environmental justice, just sustainabilities, ecological gentrification, Right to the City, shrinking cities, and the urban commons. I've attempted to draw a thread to interweave this disparate patchwork of ideas, hoping that their interrelatedness will be apparent to the reader. It is my wish that as this theoretical body matures that these respective subcomponents will become increasingly congealed and that it will be further expanded by relevant concepts which I either have omitted or am ignorant of. In the following part, I will attempt to integrate the insights of each of these concepts into a more nuts-and-bolts exploration of urban ecosystem justice on an applied level.

Works cited

Agyeman, J. (2005). *Sustainable communities and the challenge of environmental justice*. New York: New York University Press.

Agyeman, J. R. (2002). Exploring the nexus: Bringing together sustainability, environmental justice and equity. *Space and Polity*, 6(1), 77–90.

Allen, B. L. (2013). Justice as measure of nongovernmental organization success in postdisaster community assistance. *Science, Technology & Human Values*, *38*(2), 224–249.

Axel-Lute, M. (2019, July 31). It doesn't matter if your neighborhood is going to eventually gentrify. *Shelterforce Magazine*.

Barnett, J. (2006). *Climate change, insecurity and injustice*. Cambridge, MA: Massachusetts Institute of Technology Press.

Blanco, H. e. (2009). Shaken, shrinking, hot, impoverished and informal: Emerging research agendas in planning. *Progress in Planning*, *72*, 195–250.

Bullard, R. D. (2008). *Dumping in Dixie: Race, class, and environmental quality*. Boulder, CO: Westview Press.

Checker, M. (2011). Wiped out by the "Greenwave": Environmental gentrification and the paradoxical politics of urban sustainability. *City & Society*, 210–229.

Clark, B. D. (2018). The bottom line: Capital's production of social inequalities and environmental degradation. *Journal of Environmental Studies and Sciences*, 8(4), 562–569.

Colding, J. a. (2013). The potential of "Urban Green Commons" in the resilience building of cities. *Ecological Economics*, 86, 156–166.

Colding, J. e. (2013). Urban green commons: Insights on urban common property systems. *Global Environmental Change*, *23*(5), 1039–1051.

Cole, L. W. (2001). *From the ground up: Environmental racism and the rise of the environmental justice movement*. New York: New York University Press.

Corburn, J. (2005). *Street science: Community knowledge and environmental health justice*. Cambridge, MA: Massachusetts Institute of Technology Press.

Council, N. E. (2004). *Ensuring risk reduction in communities with multiple stressors: Environmental justice and cumulative risks/impacts*. Retrieved from www.epa.gov/sites/production/files/2015-04/documents/ensuringriskreducationnejac.

Curran, W. a. (2012). Just green enough: Contesting environmental gentrification in Greenpoint, Brooklyn. *Local Environment*, *17*(9), 1027–1042.

Daly, H. E. (2011). *Ecological economics: Principles and applications*. Washington, DC: Island Press.

Davis, M. a. (2011). *Evil paradises: Dreamworlds of neoliberalism*. New York: The New Press.

Fainstein, S. S. (2010). *The just city*. Ithaca, NY: Cornell University Press.

Fortun, K. (2011). Afterword: Working "faultlines". In *Technoscience and environmental justice: Expert cultures in a grassroots movement* (p. 249). Cambridge, MA: Massachusetts Institute of Technology Press.

Frickel, S. (2011). Who are the experts of environmental health justice? In *Technoscience and environmental justice: Expert cultures in a grassroots movement* (pp. 21–39). Cambridge, MA: Massachusetts Institute of Technology Press.

Frumkin, H. e. (2005). *Environmental health: From global to local* (Vol. 11). Hoboken, NJ: John Wiley & Sons.

Gidwani, V. a. (2011). Urban commons. *Economic and Political Weekly*, 46(50), 42–43.

Gould, K. A. (2016). *Green gentrification: Urban sustainability and the struggle for environmental justice*. New York: Routledge.

Gunderson, L. H., & Holling, C. S. (2002). *Panarchy: Understanding transformations in human and natural systems*. Washington, DC: Island Press.

Hardin, G. (2009). The tragedy of the commons. *Journal of Natural Resources Policy Research, 1*(3), 243–253.

Harding, S. G. (1991). *Whose science? Whose knowledge? Thinking from women's lives*. Ithaca, NY: Cornell University Press.

Harvey, D. (2008). The right to the city. In *The city reader* (Vol. 6, pp. 23–40). Basel, Switzerland: Birkhäuser.

Harvey, D. (2010). *Social justice and the city* (Vol. 1). Athens: University of Georgia Press.

Herrmann, D. L., Schwarz, K., Shuster, W. D., Berland, A., Chaffin, B. C., Garmestani, A. S., & Hopton, M. E. (2016). Ecology for the shrinking city. *BioScience*, 66(11), 965–973.

Hess, D. J. (2009). *Localist movements in a global economy: Sustainability, justice, and urban development in the United States*. Cambridge, MA: Massachusetts Institute of Technology Press.

Hodson, M. a. (2009). Urban ecological security: A new urban paradigm? *International Journal of Urban and Regional Research, 31*(1), 193–215.

Holifield, R. M. (2009). Introduction spaces of environmental justice: Frameworks for critical engagement. *Antipode, 41*(4), 591–612.

Hollander, J. B. (2011). The bounds of smart decline: A foundational theory for planning shrinking cities. *Housing Policy Debate, 21*(3), 349–367.

Hughes, S. a. (2020). Just urban transitions: Toward a research agenda. *Wiley Interdisciplinary Reviews: Climate Change, 11*(3).

Lefebvre, H. (1972). *Le Droit à la Ville*. Paris: Anthropos.

Marcuse, P. (2012). Whose right(s) to what city? In *Cities for people, not for profit: Critical urban theory and the right to the city* (pp. 24–41). New York: Routledge.

Marcuse, P. e. (2009). *Searching for the just city: Debates in urban theory and practice*. London: Routledge.

Miraftab, F. (2004). Making neo-liberal governance: The disempowering work of empowerment. *International Planning Studies*, 9(4), 239–259.

Molotch, H. (1976). The city as a growth machine: Toward a political economy of place. *American Journal of Sociology*, 82(2), 309–332.

Nagendra, H. a. (2014). Applying the social-ecological system framework to the diagnosis of urban lake commons in Bangalore, India. *Ecology and Society, 19*(2), 67.

Nassauer, J. I. (2014). Urban vacancy and land use legacies: A frontier for urban ecological research, design, and planning. *Landscape and Urban Planning, 125*, 245–253.

Ostrom, E. (2015). *Governing the commons*. Cambridge: Cambridge University Press.

Oswalt, T. R. (2006). *Atlas of shrinking cities*. Ostfildern: Hatje Crantz.

Peck, J. (2015). *Austerity urbanism: The neoliberal crisis of American cities*. New York: Rosa Luxemburg Stiftung.

Pelling, M., & Manuel-Navarrete, D. (2011). From resilience to transformation: The adaptive cycle in two Mexican urban centers. *Ecology and Society, 16*(2).

Pellow, D. N. (2017). *What is critical environmental justice?* Hoboken, NJ: John Wiley & Sons.

Popper, D. a. (2010). Smart decline in post-carbon cities: The Buffalo Commons Meets, Buffalo, New York. In *The post-carbon reader: Managing the 21st century's sustainability crises*. Retrieved from www.library.uniteddiversity.coop/Transition_Relocalisatio

Raworth, K. (2012). A safe and just space for humanity: Can we live within the doughnut. *Oxfam Policy and Practice: Climate Change and Resilience*, 8(1).

Rhodes, J. a. (2013). Shrinking "Smart"? Urban redevelopment and shrinkage in Youngstown, Ohio. *Urban Geography*, *34*(3), 305–326.

Robbins, P. (2011). What is political ecology? The hatchet and the seed. In *Political ecology: A critical introduction* (Chapter 1). Hoboken, NJ: John Wiley and Sons.

RTC. (2017, February 3). *Mission and history*. Retrieved from Right to the City, www.righttothecity.org

Schatz, L. (2013). Decline-oriented urban governance in Youngstown, Ohio. In *The city after abandonment* (p. 87). Philadelphia, PA: University of Pennsylvania Press.

Siefkes, C. (2009). The commons of the future. *Z-NET*. Retrieved November 23, 2016, from www.zcomm.org/znetarticle/the-commons-of-the-future-by-christian-siefkes/

Steele, W. e. (2012). Planning the climate-just city. *International Planning Studies*, *17*(1), 67–83.

Swyngedouw, E. (2009). The antinomies of the postpolitical city: In search of a democratic politics of environmental production. *International Journal of Urban and Regional Research*, *33*(3), 601–620.

Swyngedouw, E. (2016). The mirage of the sustainable "Smart" city planetary urbanization and the spectre of combined and uneven apocalypse. In *Cities in the 21st century* (Chapter 11, p. 134). New York: Routledge.

Sze, J. e. (2018). *Sustainability: Approaches to environmental justice and social power*. New York: New York University Press.

Tumber, C. (2011). *Small, gritty, and green: The promise of America's smaller industrial cities in a low-carbon world*. Cambridge, MA: Massachusetts Institute of Technology Press.

USEPA. (2020). Environmental justice. *United States Environmental Protection Agency*. Retrieved from www.epa.gov/environmentaljustice

Wiechman, T. (2007). Between spectacular projects and pragmatic deconstruction. Paper presented at Conference on *The future of shrinking cities: Problems, patterns, and strategies of urban transformation in a global context*. Retrieved from www.rop.tu-dortmund.de/c

Wiechmann, T. a. (2012). Urban shrinkage in Germany and the USA: A comparison of transformation patterns and local strategies. *International Journal of Urban and Regional Research*, *36*(2), 261–280.

Wolch, J. R. (2014). Urban green space, public health, and environmental justice: The challenge of making cities "just green enough". *Landscape and Urban Planning*, *125*, 234–244.

Part II

Urban ecosystem justice – urban ecosystem justice applications

3 Urban soil justice

Ecoremediation versus the toxisphere

Living soils are resilient soils. From growing nutritious food, to purifying water, and for absorbing carbon from the atmosphere, healthy soils are critical for the survival of humanity. Urban soils, however, have been widely abused. Paved over, compacted, degraded, and poisoned, the poor condition of urban soils is a significant limiting factor to achieving urban ecosystem justice. Considering this, it comes as no surprise then when many urban residents aren't aware of the difference between living soil and lifeless dirt. Worse still, because of soil contamination many city dwellers rightfully have a fear of touching or growing food in the soils in their communities, contributing to ecological alienation overall. One of the most important objectives of urban ecosystem justice is to teach the public to appreciate the amazing biological richness, complexity, and value of living soil: providing them with a tool kit of skills useful for building soil where there is none, revitalizing it when it is damaged, and detoxifying it when it is contaminated. This chapter provides both practical and theoretical tools for furthering further social justice in the urban ecosystem and for promoting a healthier and more reciprocal relationship between people and city soils.

To begin, it's necessary to have a better understanding of soil itself and the science behind it.

What is soil?

Soil is a living system. An intricate web of sand, silt, clay, minerals, water, air, organic matter, and life. Healthy soils are teeming with billions of organisms, engaged in an astounding variety of ecological exchanges and co-creative processes. This sphere of interaction between the earth, the atmosphere, water, and life is scientifically referred to as the "pedosphere". The levels of biological diversity and richness in soil are incredible. At the risk of seeming reductive by breaking down this amazing complex whole into its individual parts, it's helpful to express this diversity in numbers. One cubic meter of soil can contain:

- Over ten trillion bacteria and archaea, representing thousands of species mostly unknown to science
- One hundred billion individual fungi and actinomycetes

DOI: 10.4324/9780367858711-4

- One hundred million algae
- One million nematodes (or roundworms, as they're also referred to)
- Tens of thousands of springtails, mites, centipedes, and millipedes
- And large numbers of earthworms, ants, lice, beetle larvae, fly larvae, spiders, and a variety of other invertebrates

(Atlas, 2015); (Liittschwager, 2012)

Soil life can also be thought to include the plants growing in the ground as well as the larger subterranean mammals such as moles, groundhogs, and rabbits. Furthermore, remembering that *humus* (meaning the top layer of organic matter in soils) and *human* share the same etymological origins, surface-dwelling humans are additionally cocreators of soil and should be thought of being part of the soil ecosystem.

While the diversity of soil life is incredible, it's important to remember that healthy soils are produced and preserved through the *complex and synergistic relationships* of the different organisms that make up the soil ecosystem. These symbioses take on a wide variety of forms:

- Plants and bacteria have coevolved a mutualistic relationship that consists of bacteria making soil minerals biologically available to plants in exchange for feasting on carbon-based glucose exudates from their roots. The sticky compounds produced by the bacteria in turn then gives soil a spongy gumminess that improves its ability to absorb and retain moisture (Van Der Heijden, 2008). Nitrogen, abundant in the earth's atmosphere but scarce in soils, is made available to plants largely through the abilities of certain free living and plant symbiotic nitrogen fixing bacteria in the soil.
- The miles of mycelial threads woven through soils by fungi literally bind them together as they in turn decompose dead matter. Mycorrhizal fungi act as auxiliary root systems for plants, accessing moisture and nutrients beyond the plant's reach in exchange for carbon. In a relationship that has been described as "beyond mutualism" mycorrhizae broker nutrient exchange between trees in a forest, allowing "mother trees" to direct resources to smaller trees growing in the shade (Bledsoe, 2014).
- Tunnels made by ecological engineering earthworms create conduits for moisture and air to penetrate the soil's depths, while the soil particles consumed and expelled by them are glued together by digestive mucuses. Soil invertebrates are engaged in an ongoing dance of birth and death, while subterranean and surface animals defecate and die in and on soil, providing nutrients to the decomposers within and below.

(Jouquet, 2006)

While this is just a snapshot of the incredible complexity of ecological exchanges occurring in soil, it serves to demonstrate that healthy soil can exist only through the processes of the multispecies community that *create it* (Puig de la Bellacasa, 2015). Defying attempts to catalogue and control it, there is a recognition among soil scientists that the complexity of soil ecosystems cannot be accurately replicated in the laboratory, with microbes changing both structurally and genetically in response to laboratory conditions. Such controlled experiments only mimic or partly explain the processes at work in soil (Tate, 1995). Although there is much that can be learned about soil through reductionist approaches, there is too little known about how soil functions as a whole, how its parts interact with each other, and to how soils respond to human and natural forces over decades and centuries (Kubiëna, 1970).

Urban soils carbon capture

The carbon used to make root exudates comes from the atmosphere, scrubbed out of the air by plants. When carbon is removed from the air, stored, and left undisturbed in the soil in the form of organic matter, it is possible to use our soils as a **carbon sink** and slow climate change. Soil-carbon farming, as it is referred to, is one of the best examples of simple citizen-driven and decentralized technologies for fighting climate chaos. While most studies have looked at the potential to do soil-carbon capture on a global scale with large tracts of farmland, there is increasing interest in the idea of using urban soils to capture carbon. Because cities often contain numerous empty tracts and vacant lots, using urban soils for carbon capture is a regenerative environmental justice strategy that promotes both healthy soils, food justice, and a more stable climate (Washbourne, 2012); (Renforth, 2011); (Vasenev, 2018).

Soils and humanity – a coevolution

As stated, good soil health is the foundation of nutritious food production and therefore of civilization itself. The relative sustainability of societies throughout history can often be attributed to how they cared for their soils. Cultures that preserved and regenerated the health of their soils by cyclically returning nutrients to the earth would persist, while those cultures that depleted them either collapsed or were forced to move elsewhere, leaving deserts in their wake (Diamond, 2005); (Hillel, 1992). When soils are abused or neglected, subject to drying, compaction, salinization, or poisoning, the living matrix inside them deteriorates. Without the organic glues binding them together, soils become further vulnerable to erosion via wind and water, leaving behind lifeless bare rock incapable of supporting plant life (Hillel, 1992). The "dust-bowl" of the 1930s serves as a recent historical reminder of what the consequences of soil abuse resemble.

Following the advent of the Green Revolution of the 1950s and the wide-scale adoption of chemically based industrial agriculture, there has been a systematic abandonment of regenerative agricultural practices and with it a forsaking of living soils. On account of this, the critical web of relationships between humans, plants, and microorganisms has been largely neglected and forgotten by modern agronomists (Tilman, 1998). Plants, when given synthetically derived exogenous fertilizers, fail to develop symbiotic partnerships with soil bacteria. When given synthetic nutrients, they are no longer reliant on soil microorganisms and will cease giving off their root exudates to feed them. By engaging exclusively in NPK (nitrogen, phosphorus, and potassium) farming, plants become deficient in micronutrients that are otherwise provided through microbial symbiosis, and in turn so do the humans that eat them. Soil is then regarded as, and becomes, inert lifeless dirt whose only function is to physically prop up a plant. Through unsustainable agricultural practices over time, crop yields decline and in a positive feedback loop of dependency and despair, farmers become increasingly reliant on chemical and synthetic fertilizer inputs, making it ever more difficult to grow food without them (Shiva, 2008). Luckily, organic farming practices once regarded as fringe are now becoming more mainstream and with them an improvement in soil health and biodiversity, both below and above ground (Birkhofer, 2008).

With an understanding of the impact that human activity has on soil and how soil health in turn affects human well-being, it becomes clear that humans are cocreators of soils and that humans are part of socially constructed soil ecosystems. In recognition of this, a call has been made for the development of the discipline of anthropedology to study human–soil relationships (Bacon, 2011); (McClintock, 2010). The importance of understanding human–soil co-genic relationships is of utmost importance in the Anthropocene. Not only do humans possess the ability to dramatically transform soils, but the evolution of human society and culture is in turn influenced by changes in soil (Richter, 2015). Soils must be capable of sustaining a near doubling of demand for global food production in the coming decades and must also continue to act as resiliency buffers, providing a host of ecosystem services including flood mitigation, biodiversity support, nutrient and contaminant filtration, disease production or suppression, and greenhouse gas absorption or emission. If soil indeed reflects “the memory of the landscape” (Targulian, 1978), then the soils of the future will tell a story to our descendants about how wisely or foolishly humanity navigated the Anthropocene.

Soils in the city

Similar to waterways, the soils in the near vicinities of cities have played an important historical role in sustaining urban functioning. It is no coincidence that many cities are in areas with some of the most fertile soils. River valleys, common sites for urban development on account of all the benefits of waterways, also frequently possess high-quality soils. Many of the organic nutrients that are washed away from higher altitudes will accumulate in these regions, and the periodic flooding of rivers will deposit nutrient-rich sediments upon the alluvial plains. Societies such as those along the Nile in ancient Egypt were able to sustain themselves for centuries through farming in river valleys. Similarly, before the widespread use of internal combustion and the beginning of the sanitary cities movement, it was common practice for night soils (human wastes) to be returned to farms in the urban periphery, sustaining a cycle of nutrients (Atkins, 2012). Even though cities have been historically built in regions with high soil fertility to support their populations, the health of soils in today's cities is rarely considered by residents or planners. As urban metabolic processes have become increasingly linear not only has the fertility of peri-urban soils been neglected, but suburban sprawl beyond city limits has transformed these once fertile fields into monocultures of modular homes and St. Augustine grass lawns.

As part of this construction process, soils rich in organic matter have been cleared in order to expose the firm mineral soils below. From the perspective of a building engineer, soils are suitable only if they are compactable – capable of being compressed into a hardened base able to support the weight of structures. Organic matter – the living, soft, glue-like material that gives healthy soils their vitality – is considered undesirable as it will decompose over time and result in the subsidence and shifting of structures. For this reason, fertile topsoils are commonly scraped from a building site before any construction begins (Egendorf, 2018).

In many instances, indigenous soils are replaced with "fill" soils which consist of a mix of materials ranging from rock, mine tailings, dredging sediments, brick, concrete, coal ash, and garbage. Anyone who has ever dug a series of holes in a city can attest to huge variability of urban soil composition. Even digging a single hole will often reveal a series of layers of soil types stacked on top of each other – both natural and human-made. A hole several feet to one side or the other can have an entirely different makeup. These layers create a fascinating historical record of a city's past and can make for a fun archaeological project. They do make it a challenging puzzle, however, to know what the natural pre-urban composition of soils consisted of. To reflect the huge variability and anthropogenic impact on city soils, urban ecologists use terms such as "technosols" (Séré, 2010) and "anthrosols" (Smith, 1980) to describe them.

Challenges of urban soils

As discussed, urban soils are frequently nonexistent, degraded, or contaminated – creating significant barriers to food production and urban ecosystem justice. This section will discuss each of these issues in detail, along with possible remedies for each.

Nonexistent soils

"Nonexistent" soils refer to soils that have been covered and sealed with impervious materials such as asphalt and concrete to create roadways and parking lots. To describe these soils as nonexistent is a bit misleading – they exist, they are just buried and starved of air, moisture, and nutrient exchange – effectively making them dead, sterile, nonliving mineral substrates. Coating soils with impervious toppings, or "sealing", has enormously negative consequences for the health of the urban watershed – by destroying soil's capacity to absorb, retain, and filter stormwater, pollutant-laden volumes of runoff overwhelm drainage systems, causing sewage overflows and toxic flooding (Scalenghe, 2009). Food production is also seriously limited on paved surfaces (unless one can survive off mosses and lichens). For these reasons, removing impervious covers from soils is an essential first step toward regenerating the health of urban ecosystems.

Parking lots and lava fields – ecological analogues

Hot, molten rock – steaming and slow moving, releasing hazardous vapors, killing everything in its path and eventually hardening into an impervious blanket after it finally cools. This describes both lava flowing out of a volcano and asphalt being dumped out of the back of a truck. Yes, clearly one is a "natural" process where the other is anthropogenic, yet they possess some ecological similarities, and studying the successional processes that follow a volcanic eruption can give us clues about how to regenerate urban ecosystems. This is an important topic as the question often arises among urban gardeners of whether it is really necessary to remove asphalt, and whether it is possible to just garden on top of it. While it may be important to remove impervious covers for the long-term health and regeneration of urban ecosystems, there may be reasons and circumstances why that can't or shouldn't be immediately done. One of these could be cost and energy – taking up asphalt is expensive and time-consuming. One may wish to just start gardening right away or may not want to invest the time and energy into revitalizing a parcel you don't have secured access to. Also, sometimes soils have been sealed as a means of encapsulating seriously contaminated soils below. If this is known to be the case, it could be a good argument for leaving impervious covers in place. Or perhaps it's a paved vacant parcel behind a secured fence that one would wish to start healing via guerilla tactics (i.e., seed bombing) (DiSalvo, 2014). In any case, one could think of the asphalt as being a bedrock layer, and just begin building soil on top of it.

Going back to the lava field analogy, soils will start to regenerate on heavily disturbed environments over time. Just observe an abandoned parking lot – cracks form on its surface that allow ruderal (disturbance loving) plants to access the pockets of soil underneath. Over time, plant roots widen these cracks and themselves die, forming more soil in the process. Ruderals, or disturbance-tolerant organisms like mosses and lichens will colonize hard surfaces, helping as well to regenerate soils (Walker, 2012). Eventually, a thin soil layer will form that will grow over time, allowing larger plant species to survive. This process can be allowed to happen at its own pace or sped up through human intervention. Twelve inches of soils will create enough soil depth to plant annual vegetables. This could be done by dumping soil on top of a lot or by building raised beds on its surface.

The drawback is that the opportunity to get the impervious layer up is lost, which would allow plant roots to penetrate deeply (essential for trees) and for water to infiltrate into the subsoil. It's always possible to go back and do this later but will result in serious disruption to whatever soil has been created since then.

Ripping up impervious covers

Tearing up asphalt is a surprisingly pleasurable activity – there is satisfaction in feeling compacted and nutrient soils breathe again after years of being sealed. However, depending on how thick the asphalt is, and how large an area it covers, this can be either a relatively simple or difficult job. Sometimes, if the asphalt is an inch thick or so, it can be pried up in large chunks using a rock bar and a pickaxe. These conditions make for easy de-paving and can be done in a couple of hours with a group of volunteers. De-paving areas where the asphalt is six inches thick or more, on the other hand, is an exhausting task in and of itself. In these situations, it may be necessary to bring in excavators to tear up a parking lot with the understanding that this one-time expenditure of fossil fuel energy will yield long-lasting benefits to urban communities. What then, should be done with the asphalt that has been ripped up? There are some companies that recycle old asphalt, using it to make road-base for new road construction elsewhere (Miliutenko, 2013). While some urban gardeners repurpose asphalt chunks to make raised beds, there seems to be some uncertainty as to whether this is a good practice or not on account of the possibility of chemicals leaching out into garden soils. Considering all options, it may be preferable to have asphalt landfilled. While having to throw anything into a landfill is unfortunate, in the absence of a better use it may be best to put asphalt back into the ground, where it came from in the first place.

Areas covered by concrete are a different story. Usually, these are old building foundation slabs and have a metal rebar lattice inside of them that makes them nearly impossible to tear out by hand. Taking them up will require a jackhammer at minimum and likely something much bigger like an excavator. Reuse potential of concrete is a bit better than asphalt, however (Figure 3.1). Concrete chunks are relatively inert and have great compressive strength, making them good for raised beds, retaining walls, and so on. Rather than removing concrete foundations, however, you might want to think about what they could be reused for. At Radix, honeybee hives have been placed on concrete pads, which make stable foundations for them and minimizes the need to mow too close to the hive, which disturbs the bees.

Degraded soils

Removing impervious covers alone, however, is insufficient for restoring soil health. As mentioned, the soils beneath parking lots and roadways are largely nonliving.

Figure 3.1 De-paving
Source: Scott Kellog

In addition to being deprived of air and water, they've also been compacted. The weight of the asphalt or concrete on top of them combined with cars and trucks repeatedly driving across have squeezed the soils flat and turned them into "hard-pan", an impervious layer of compressed soil. Even more compressed are soils that formerly had buildings on top of them – the weight of the structure above combined with the deliberate pre-compaction of the soils before construction will result in some seriously squeezed earth. These compacted, dead soils have an absorptive capacity that is only slightly better than that of asphalt and will need lots of healing before they are once again capable of supporting much life (Stoessel, 2018).

Because the particles in compacted soil have been squeezed together, any air or moisture that might have otherwise been between them has been forced out. This makes it very difficult for water, plant roots, worms, or just about anything else to penetrate through the soil column. In order to convert dead dirt into living soil, it will be necessary to loosen it so that air and moisture can once again flow through it.

One way to test to see if soils are compacted is to try and shove a pencil down in the ground. If it can't be forced all the way in, then the soils are likely compacted. Relieving compaction can be a challenging task. Mechanically, this is often done with rototillers – machines that turn over the top few inches of soil. This causes significant disturbance to the soil profile, however, and may not be possible in soils that are heavily compressed. There are, fortunately, simpler methods of relieving soil compaction that rely on plants. Through a process called biological tillage, it is possible to break up soil layers with plant roots, increasing their organic matter content and making it easier for worms and moisture to move more deeply into soil layers. One plant that is especially effective at doing this is the forage radish, also known as the daikon radish. It forces a taproot down through compacted soils, breaking them apart as it grows to an inch or more in diameter. Allowed to remain in the soil, it will eventually die and rot, enriching the ground with its nutrients and creating a conduit for moisture and oxygen. For maximum effect, forage radishes should be seeded early in the growing season so that they can reach their full size before dying in the winter freeze (Chen, 2009).

Once soils have been relieved of compaction, it becomes possible to start bringing them back to life. While soil microorganisms will eventually recolonize soils, the process can be accelerated by introducing microbial inoculants and by better improving soil conditions. Both goals can be accomplished by applying compost to soils. While both compost and composting are discussed further in the chapter on waste, their benefits cannot be overstated and are worth reiterating here. Finished compost is exploding with microbial life and can hugely jumpstart the process of soil revitalization. Even by spreading it across the surface of soils or by lightly tilling it in, microbes will be reintroduced in droves, washed downwards into the soil by the rains. Once they are established, and if they survive, the process of soil regeneration will begin (Ohsowski, 2012). These microbes act as the interface between living and nonliving matter. By literally eating rocks, they convert inorganic minerals into plant nutrients. Furthermore, the bodies of these microbes, both living and dead, and the compounds they produce are what gives healthy soil its crumbly, moist, sponge-like texture – measurable in what's called organic matter content, or OM. High organic matter content in soils is essential for both moisture retention and for proper drainage, both critical for ecosystem functioning on multiple levels. On the garden scale, high OM helps retain water in sandy soils and helps drain it away when clay content is high. Healthy soil should have a 40–60 percent moisture content – about that of a wrung-out sponge – high OM helps maintain this balance. Logically then, plants growing in high-OM soils require less watering, as stored water is made available to their roots. On a citywide and regional scale, moisture retaining soils help to mitigate flash flooding and help to buffer against droughts. In this sense, building healthy soils is an important part of a community's defense against climate volatility (Gittleman, 2017).

Finished compost, which is pH neutral, is also ideal for moderating either alkaline or acidic urban soils. Urban soils can frequently be calcareous – meaning they contain large amounts of calcium carbonate or lime (typically from concrete residues) which makes them alkaline. Adding compost can help to balance out this alkalinity. Having a balanced pH is important not only for healthy plant growth but also for immobilizing pollutants, as described later in this chapter.

While the systematic neglect, abuse, degradation, compaction, covering, and all together erasure of urban soils have contributed to people's alienation from them, another significant factor that continues to generate ignorance and fear of soil in urban residents is the issue of soil contamination.

Soil contamination

We live in a toxic world. From heavy metals to combustion by-products to venomous animals, organisms and ecosystems have evolved fascinating methods of avoiding, degrading, removing, or otherwise coping with poisonous substances. While some species can adapt to toxic conditions and survive, others do not and end up going extinct. Sometimes, toxins that are deadly to one species become essential for another (Monosson, 2012). While it could be argued that toxicity is to some degree "natural" and part of the ecological adaptive cycle of growth and release – there are limits – and through modern industrial processes, these limits have been exceeded. If we care about the health and survival of humanity and of most other species on the planet, we must discontinue the reckless release of toxins in our global environment and begin the process of cleaning up the mess that has been created. This section will discuss pollution in cities, ways to protect communities from it, and adaptive methods for detoxifying urban ecosystems.

Contaminants of concern

Very broadly speaking, there are two general categories we can put contaminants in to. The first is elemental heavy metals, chiefly lead, arsenic, cadmium, mercury, and chromium. They exist in nature, albeit often in deposits deep in the earth, bound up in rock and isolated from the biosphere in anything more than trace quantities. Through extractive mining processes, these metals have been brought to the surface of the planet and diffused through the ecosystem. They now exist in concentrations greater than what we have evolved to be able to deal with. High concentrations of toxic metals are dangerous to most living things, resulting in a host of health maladies (Storelli, 2008).

Due to both its ubiquity and toxicity in urban ecosystems, lead (Pb) is the most concerning of all metals. There are many sources of lead in urban environments, but the two main ones are old lead-based house paint and auto exhaust. While both have been prohibited in the United States since the 1970s, their legacies remain with us to this day. Often, the highest concentrations of lead are found along the drip lines of homes or along roadways where leaded exhaust was discharged. Lead is a neurotoxin. It readily binds with brain tissues and causes developmental impairments in children (Mielke, 2012). The following sections will discuss methods for dealing with lead contamination.

The second broad category of toxins are what are called organic pollutants. These consist of compounds made up of molecular chains and can include hydrocarbons, pesticides, volatile organic compounds (VOCs), polycyclic aromatic hydrocarbons (PAHs), chlorinated solvents, dioxins, PCBs, and so on. There are over 80,000 of these chemicals in use in the United States, most of which have not been adequately tested for their effects on human health (let alone their synergistic toxicity – or how they can become even more toxic when blended together). While some of these readily break down in the environment, others are recalcitrant – meaning they are resistant to degradation. The longest lived of these comprise a group of toxins called persistent organic pollutants or POPs (Bernes, 1998).

Toxic cities

Cities historically have served as centers of industry: manufacturing, smelting, and refining businesses have been commonly located in urban areas. Because of this, many pollutants produced as by-products of these industries over the past two centuries have been released into urban ecosystems (Hough, 2004). Some of these pollutants are "diffuse", meaning they are emitted into the air through smokestacks and are spread broadly over an area. Others are "point source", which basically means they were dumped and can be found in high concentrations in specific places. Although air and water are regularly polluted, the contaminants they bear eventually fall out of suspension and concentrate in soils, making urban soils a toxic substance sink. For this reason, it makes sense to include a general discussion of toxicity in the chapter dedicated to soils.

As an issue, urban soil contamination has gained an amount of attention recently as interest in urban agriculture and community gardening has increased. Because gardeners come into direct contact with potentially contaminated soils, inhaling their dusts, and growing food in them, they are potentially at a higher risk of harmful exposure (Cao, 2010); (Prasad, 2000); (Alloway, 2004). Low-income urban populations are at particular risk, as polluting industries are more likely to be situated within low-income neighborhoods. Additionally, negligent landowners are less likely to have carried out lead paint remediation in poorer neighborhoods, which may continue to impact the soils and the health of communities to this day (Pirkle, 1998).

One of the most frustrating aspects of soil contamination (and of toxic exposures in general) is the huge amount of uncertainty surrounding it, particularly regarding health concerns and whether activities like gardening are "safe". As there are many thousands of known toxic substances, comprehensively testing for all of these is incredibly expensive. Furthermore, there are tremendously complicated interactions between contaminants, soils, plants, and people, creating a huge number of open questions about what toxins may be present and where they end up (Vidali, 2001).

Some of these key questions for creating urban soil justice are:

(1) What contaminants are in the soil?
(2) Are those contaminants getting into human bodies, either through contact with the soil or by eating plants grown in it?
(3) If they are entering our bodies, what effects then are they having on our health?

Because risk analysis frameworks will never be able to take all complexities into account nor be accessible to everyone, these questions often go unanswered.

Consequently, many people act in a precautionary way and will choose not to garden in urban soils.

While there are always potential risks, there are also many substantial personal benefits to urban agriculture that should be considered (Wakefield, 2007). These include improved nutrition, exercise, time outdoors, mental stimulation, and social connections. Going beyond just the personal health benefits, the broader social, economic, and ecological benefits of urban agriculture are tremendous. From creating local food economies, combatting food apartheid, gardening education, climate mitigation, biodiversity enhancement, and overall resilience building, there is no question that urban agriculture is a critical component of sustainable city systems and urban ecosystem justice overall.

Risk-reducing best practice strategies

There are several simple risk-reducing practices that can be used by urban gardeners to reduce their chance of exposure to soilborne toxins (USEPA, 2009). These strategies, considered "best practices", make good sense to employ when faced with toxic uncertainty.

Some of these include:

- **Removing shoes before entering a home after being in the garden**
  Just by doing this, the amount of potentially lead-contaminated soil being tracked into the home can be significantly reduced. This is especially important if young children are in the home who are crawling on floors and putting their hands in and out of their mouths – remember lead poisoning is a much greater concern for children than adults.
- **Washing**
  Washing both hands and vegetables is important for reducing exposure to soil toxins. This is especially the case for root crops (carrots, potatoes, beets, radishes, etc.) that have grown in the soil. It is recommended that either soap or water mixed with 1 percent vinegar be used for washing vegetables.
- **Raised beds**
  If soil contamination is a concern in an area where you want to garden, a smart and easy solution is to construct raised beds. Raised beds are basically elevated and contain mounds of soil around six inches deep or more. They are most frequently constructed from wooden boxes framed with lumber, but can also be created with stones, logs, bricks, or concrete chunks. The basic idea is that you are creating a garden bed on top of ground with imported "clean" soil. As most annual garden vegetables do not have a rooting depth of greater than six inches, the raised bed prevents plants roots from accessing potentially contaminated soil. As an added precaution, the bed may be lined on the bottom with landscaping fabric, a material that permits water to pass through yet blocks plant roots.

Raised bed considerations and concerns

Although purchasing soil and box materials are expensive, the cost can be justified in an urban environment as the "solar real estate" is of greater value than the soil – particularly if it's only a few beds. What this means is that having access to a space that receives enough sunlight to support vegetable production is the most important thing for gardening. Frequently, vacant parcels are shaded by trees and buildings to such an extent that vegetable gardening is impossible – it's much easier to import or amend soil than to knock down neighboring structures. Once you pass from small-scale gardening into larger scale farming, the value of good indigenous soil becomes more important.

Much of the soil that is sold by landscaping companies is collected by scraping the topsoil off development sites prior to construction. It's possible that this topsoil may be more contaminated than the soils it's meant to cover. Always ask landscaping companies if they screen their soils for lead. Another concern is about the long-term sustainability of the practice of harvesting topsoil. Any city has a theoretical "soil shed", the area surrounding a city from which soils are collected from. In keeping with the goal of having cities function in a more cyclical and less linear way, it's important to not to vacuum the soil from rural farms into cities. Responsible landscaping companies will "grow" their soils, mixing in composted organics with parent material and not "flay" the neighboring fields of fertility.

- **Ground covers**
  Establishing grounds covers is an effective way of preventing contaminated soils from drying out and becoming dust-borne or from being splashed by rain or tracked by foot into uncontaminated areas. A vegetated cover (i.e., grass) maintained over soils can help create an effective means of protecting soils from erosion and drying. However, in high-traffic areas, such as the rows in between garden beds, it can be difficult to keep a ground cover alive. Other materials that can be used in these spaces include wood chips, gravel, paver stones, burlap bags, or cardboard. Any of these will help to keep contaminated soils from drying out or being spread. Do keep in mind, however, that materials like wood chips will degrade and will need to be reapplied at least on an annual basis.
- **High organic matter content and neutral pH**
  Lead is highly lipophilic (lipid=fat, philic=loving). Therefore, it bonds readily to receptor sites in human brains, which are themselves about 60 percent fat (Lidsky, 2003). Luckily however, lead's lipophilic properties also lend to it binding to organic compounds in compost amendments. When bound with

compost, lead's bioavailability is reduced, which means it is less likely to be taken up by plants or to cause neurological damage if ingested. Furthermore, adding compost to soils neutralizes their pH, causing toxic metals to be less soluble and unlikely to be absorbed by plant tissues. Following a similar logic, a nutritious diet (particularly in calcium and iron) will help to protect the body from lead poisoning, as these two minerals compete with lead for uptake in both the gut and the brain (Sargent, 1994).

- **Aerial deposition barrier**
  Even if a garden bed is built with clean soil, or is placed on an uncontaminated site, it is still important to protect soils from dust-borne contaminants blowing in from off-site. These can originate from uncovered soil in a neighboring yard with flaking lead-based house paint or could be from street dust – contaminated by degrading brake pads, diesel particulates, and wearing tires. One way to protect a site is to create a depositional barrier, a row of nonedible plants grown around the periphery of a property. The leaves of these plants will create a surface for airborne pollutants to settle on, blocking some of them from moving further into a property where they could contaminate garden soils. Barriers make good sense if part of a property is facing a heavily trafficked street (Antisari, 2015).

These best management practices are effective at protecting gardeners from direct exposure to contaminated soils. They offer simple and practical methods of creating first line of defense against urban ecological contaminants.

Bioremediation – proactive solutions to soil contamination

What would it mean, however, to go beyond just a reactive, defensive response to pollution? Is it possible to "go on the offensive", so to say with environmental toxins, and intervene in such a way where the risk could be permanently eliminated, resulting in healthy soils for the generations to come?

In order to address soil contamination in a meaningfully long-lasting way, techniques need to be employed that are capable of degrading contaminants into their innocuous components or permanently sequestering them so that they will pose no risk to future gardeners. In keeping with an ethic that values soil as both precious and vulnerable, such long-term strategies are required. Bioremediation may be one possible method of achieving these goals.

Bioremediation is the process of using the biological properties of naturally occurring organisms – chiefly microorganisms (bacteria), fungi, worms, and plants – to degrade, immobilize, or sequester both organic and elemental environmental toxins. Bioremediation offers a more sustainable alternative to the conventional practice of "dig and dump" where contaminated soils are removed from a site and landfilled elsewhere. Although dig and dump methods rapidly fulfil regulatory obligations for site cleanup, they in no way permanently address the issue of soil contamination and just move the problem from one locale to another. As there is a growing recognition that there is no such thing as "away", bioremediation provides a method of soil remediation that can be performed "in situ" with minimally disruptive techniques (Cookson Jr, 1995). While bioremediation operates on a longer timescale than conventional cleanup methods (projects can take years to complete), it treats soils like the precious resources that they are, leaving them available to future generations.

Community-based bioremediation

While bioremediation has the *potential* to be carried out in way that makes it simple, affordable, and a tool that citizens can use for taking proactive measures against environmental contamination, it is most commonly not. Conventionally, bioremediation is employed by professional engineers who have extensive backgrounds in the environmental sciences. They employ a wide range of proprietary techniques and technologies in order to promote the biodegradation of pollutants in the environment and rely upon access to sophisticated monitoring equipment to verify the effectiveness of their practice. They frequently also have clientele with significant financial resources who can afford their costly services. Increasingly, in the interests of maximizing profits, trademarked blends of microorganisms, some even genetically modified, are being used by bioremediation engineers (Drobník, 1999). While bioremediation shows great promise, and while there have been recent efforts to develop "green" or low-carbon bioremediation technologies (Hou, 2014), there are significant barriers that impede its wide-scale use by ordinary citizens and environmental justice communities. While it could be argued that the potentially hazardous work of bioremediation is best done by trained professionals, many of the communities most threatened by toxicity (such as low-income urban and rural communities, and residents of the Global South) cannot afford professional engineering services. Furthermore, as these communities' survival often depends upon farming, fishing, and other activities putting them in close contact with soil and water, there is an immediate imperative for them to proactively address environmental contamination. It should also be stated that in widespread disasters, such as hurricanes, the ability of governments or private companies to respond to the multiple ensuing toxic threats is extremely limited (Santos-Lozada, 2018), and citizen-based responses are required to protect the public health.

In response, a grassroots movement has emerged that desires to empower communities with the tools and information necessary to address residual industrial toxicity in their ecosystems. Focusing on methods that are simple and affordable, this movement seeks to remove the barriers of cost and technical expertise that may be otherwise prohibitive. This movement envisions community-based bioremediation as a form of citizen science, where the tools for data collection, soil analysis, and degradation of toxins are put into the hands of marginalized community members historically most disproportionately affected by toxicity (Darwish, 2013); (Kellogg, 2016).

Citizen bioremediation provides further benefits by constructing mutually symbiotic partnerships between humans and non-humans: by providing microbes, fungi, plants, worms with ideal environmental conditions and a food source, humans are in return given detoxified and nutritionally enhanced soils. This human–microbe symbiosis works to promote healthier urban ecosystems overall. Furthermore, by becoming familiar with basic bioremediation techniques, residents can train one another in these methods as the microbial cultures used in the process can be maintained and shared, similar to how a traditional sourdough starter is passed on between families. In this sense, the value of these microbial cultures and bioremediation know-how can be continually circulated within a community. This justice-centered approach is markedly different from the conventional practice of hiring experts to fix problems at great expense and leaving communities with no lasting tools for continued empowerment (Eglash, 2014).

Turning off the tap

Community-based bioremediation must be thought of as a component of a multiple-part strategy of combating toxicity – explicitly incorporated into a larger critique of the global production and distribution of toxic materials and be centered from an environmental justice perspective. Too often bioremediation is used as a justification, or an excuse, for the continued contamination of the environment: corporations pollute, engineering firms clean up the mess, and exposed communities suffer. Small-scale remediation efforts are insufficient to counter the worldwide manufacture of toxins at its current rate. For this reason, as the mess is cleaned up, we also need to be working to turn off the tap. Furthermore, corporate polluters must be held accountable for cleanup and reparations to the environments and communities who have been harmed by their actions. There also needs to be a shift/return to using materials that are nontoxic and that can ultimately be composted into organic soil. Lastly, community bioremediation advocates need to insist on the application of the precautionary principle to chemical manufacturing that materials must be proven to be safe before they are released into the environment or into bodies (O'Riordan, 2013).

No magic bullet

Environmental contamination is an incredibly complicated and genuinely wicked sustainability issue. The interactions between different chemicals and metals in various environmental media and how they interact with ecosystems and biology are terribly complex, resulting in considerable ignorance in the scientific community. As such, it is frequently difficult to know how pollutants will degrade under real-world conditions. Many of the bioremediation studies that have been carried out have taken place in controlled laboratory settings with precisely measured homogenous conditions. Under these circumstances, x species of organism may have effectively degraded × contaminant. Reproducing the same results in a highly heterogeneous and varied environment such as a vacant lot is wildly different – the leap from laboratory to the field is a large one in science dominated by reductionist methodologies. What this means is that bioremediation is a complex process with multiple variables, there is no guarantee that it will ever work, and it is difficult to distill it down into a simple set of DIY practices that can work universally against any form of contamination. In short, there's no single magic bullet that will address every toxic issue.

Diversity and non-sterility

In community-based bioremediation, "wild cultivation" methods are preferred over sterile lab techniques because most of the world's population will never have access to sterile laboratory conditions. For any method, then, to have a wide-ranging impact it must not be contingent on having access to sophisticated infrastructure. Microbiological diversity is also incredibly important. Rarely is any single microorganism capable of fully degrading a toxic compound. Frequently, a consortium of microbes is needed that can degrade the various "daughter compounds" that are produced as by-products of biodegradation (some of which can be more toxic than the original substance!). These co-metabolic processes produce an emergent and synergistic effect that can more effectively degrade a chemical in its entirety (Chemlal, 2012). To give an example, in a handful of worm compost there are likely billions of microorganisms representing an incredible diversity of species (many of which cannot even be lab-cultured or identified) (Rondon, 1999); (Daniel, 2004). The assumption is that out of that incredible diversity, the necessary organisms to degrade certain toxins are likely present and could be used as an inoculum in a bioremediation intervention. This sort of nonproprietary approach to bioremediation may be less attractive to engineering firms with its reduced profit potential but would be of greater interest to citizen's groups and activists on account of its decentralized and open source nature, replicability, and transferability.

Risk and safety

As there are inherent risks involved anytime a person comes into contact with toxic materials, it must be emphasized that personal safety takes priority over any cleanup endeavor. There is a fine line between citizen empowerment and endangerment, and it could be unknowingly crossed. Again then, would it not be better to err on the side of caution and leave bioremediation to experts who understand the risks? In my opinion, concerns over risk should not be used to shut down discussion and careful experimentation with bioremediation technologies. To argue otherwise is equivalent to saying that people should not be taught how to fish because they might possibly fall in the water and drown.

On account of this uncertainty, however, this chapter primarily provides just an exploration of the *idea* of bioremediation as a tool usable to promote urban ecosystem justice, not necessarily an in-depth technical review of bioremediation practices themselves (additional resources on this are provided at the end). Moreover, it's a way of thinking about community-based bioremediation as a different kind of science than the highly controlled, risk-averse, and reductionist variety that is commonly practiced. In many of the scenarios where community-based bioremediation would be carried out, uncertainty and stakes are both great, and levels of control are low – circumstances sometimes referred to as "post-normal" (Funtowicz, 1999). Currently, inherently reductive models for doing this type of science are poorly developed but need to be improved as complex problems such as toxicity and climate change continue to unfold.

In summary, qualities of community-based bioremediation are:

(1) Affordability
(2) Simplicity
(3) Utilizing naturally occurring and locally available organisms
(4) Preferentially employing non-sterile processes
(5) Situated with an environmental justice context
(6) "Open source" – information and data freely shared among practitioners

How it works

Remember that there are two general classes of environmental pollutants: organic and elemental, and bioremediation techniques will differ for each.

Bioremediation of organic pollutants

Organic pollutants consisting of molecular chains of various atoms can *in theory* be broken down into simpler, safer components. This biodegradation occurs in nature on a regular basis, especially with compounds such as crude oil. Crude oil is a natural substance – produced from the bodies of ancient organisms buried deep in the earth and transformed into black ooze through tremendous heat and pressure. While mostly brought to the surface with mechanical pumps, crude oil does frequently make its way through the crust in small fissures and cracks called "seeps" – natural oil seeps account for around one half of all the oil spilled in the world's ocean each year (NOAA). On account of this, it makes sense that bacteria, which evolve so rapidly due to horizontal gene transfer, have figured out how to break oil down and use it as a food source (Margulis, 2009). This ability is widespread among the bacterial domain in both terrestrial and aquatic environments, with an estimated 20 percent of soil microorganism possessing the ability to degrade crude oil (Cookson Jr, 1995). How do they do this? Crude oil is a hydrocarbon, a molecular chain consisting of, as the name suggests, hydrogen and carbon. Stated very simply, microorganisms produce enzymes that break the bonds between the hydrogen and carbon atoms, consuming the carbon as an energy source and off-gassing the hydrogen.

Sometimes, however, because of an event like an oil spill, an ecosystem's microbial cleanup capacity will be overwhelmed. Or other times contaminants may be present in environments like anoxic river sediments where microbiological activity is low, and rates of biodegradation are slow or zero (Perelo, 2010). This is where human intervention becomes necessary to speed the bioremediation process along. This could just involve introducing microorganisms in places when their populations are low (like a recently de-paved parking lot). Other times it may involve adding nutrients, carbon, oxygen, or moisture. In any instance, it's a process of getting the right organism in place that has the necessary enzymes to degrade the pollutant and creating the ideal conditions for them to do their work. It's possible for this to be partially achieved simply through good organic gardening practices: adding compost and nutrients to soils, improving air and moisture infiltration, encouraging microbial diversity, promoting earthworm colonization – these are all actions that will improve soil and microbial health overall, which will hasten the degradation of some organic toxins in the soil. This is part of the logic behind the three-year waiting period required of farms transitioning from conventional to organic – that many of the synthetic chemicals previously used will have degraded in this time (Menalled, 2009).

In other scenarios, however, more deliberate, focused, and intentional actions need to be taken.

So as a hypothetical example, let's say someone changed their oil from their car in a vacant lot and spilled oil across the ground. As it has been discussed, that would break down over time through the processes of the indigenous microorganisms on the site. However, through a bioremediation intervention, this degradation could be accelerated by repeated applications of compost tea, tilling of the soil, and mixing in compost and wood chips, shortening the rate of degradation from years to possibly months or weeks.

Decentralized and rapid-response bioremediation protocols need to be developed for use in the aftermath of catastrophic flooding scenarios when a toxic mixture of biological (sewage) and chemical (hydrocarbons) contaminants blanket an area after floodwaters recede. Conventionally, private engineering firms will be contracted (by those who can afford their services) to remediate spill contaminants. As these disaster situations will occur with greater frequency and intensity as a consequence of an increasingly volatile planetary climate, and as it will be primarily low-income communities of color who will bear the brunt of this toxic ecological exposure, and as federal support for relief in such circumstances cannot be relied on in the current political climate, the services of professional engineering firms are inadequate to address the widespread devastation and enormous need that results following disasters. For this reason, appropriate bioremediation strategies that are simple, affordable, and can be carried out utilizing locally available materials need to be employed in these situations.

The next section will describe in more detail a handful of bioremediation practices appropriate for small-scale, experimental, community-based trials. These can serve as starting points for interested activists and work well as fun educational projects. Again, it will be necessary for anyone looking to take these technologies beyond micro-scale applications to undergo significantly more study and learning – the systems described here are only intended to be a starting point.

Compost tea

Compost tea, or vermitea, is a liquid culture of beneficial microorganisms (chiefly bacteria, fungi, protozoa, and nematodes) that is used frequently in organic gardening practices as a microbial inoculant, fertilizer, and foliar spray. It is made by placing a small amount of (typically) worm compost in a container of non-chlorinated water, adding nutrients (commonly molasses, fish hydrolase, humic acid, rock dust, etc.) and aerating the mixture for 24–36 hours at room temperature. When the microorganisms in the compost receive water, nutrients, and oxygen, they proliferate exponentially and can then be added to soils.

While most studies on compost tea have examined its effectiveness in plant disease suppression and agricultural applications, it has been posited that compost tea could additionally be used as a tool for bioremediation because its use increases microbial activity and enzyme production in soils (Jonsson & Haller, 2014). While a few studies have provided some preliminary evidence for this (Haller, 2016; Chiang, 2013), others have not been as promising (Rejón et al., 2013). Generally, on account of the lack of corporate interest in funding studies of decentralized technologies with minimal profit potential, compost tea bioremediation remains in an area of "undone science" (Hess, 2016). However, as professional engineering firms utilize a liquid-based microbial culture somewhat like compost tea (albeit with a proprietary blend of organisms) for degrading organic and biological pollutants in a process called bioaugmentation or biostimulation, it's not unreasonable to think compost teas could be used similarly. This, the promising preliminary results, and its great potential for widespread application warrant further studies of compost tea's efficacy in bioremediation.

Instructions for brewing compost tea: The following is a simple description about how to brew compost tea on a very small level. Once you become comfortable with this process it can be scaled up to produce larger quantities. Note that larger amounts will require more powerful air compressors, which will increase the capital costs of the system.

What you will need:

- One five-gallon bucket
- Four gallons of non-chlorinated water (collected rainwater is fine)
- One small aquarium air pump
- A handful of finished compost
- A spoonful of molasses

What to do:

Step 1: mix water, compost, and molasses in the bucket.

Step 2: aerate the mixture with the aquarium pump for 24–36 hours.

Step 3: remove air pump and apply finished tea to soil within four hours. Note that compost tea has no shelf life and must be applied to soil quickly after aeration has stopped. Failure to do so can result in the tea becoming anaerobic and smelly.

There are many recipes and methods for producing compost tea (Figures 3.2 and 3.3). The system described here is a simple and "entry level" way to gain familiarity with the process of liquid-based microbial culturing.

Figure 3.2 Applying Compost Tea

Source: Scott Kellogg

Figure 3.3 Brewing Compost Tea

Source: Scott Kellogg

Spent mushroom compost

Another appropriate remediation technique is the application of spent mushroom compost to contaminated soils. With this method, the by-product of mushroom production is used, thereby making it possible to regenerate the health of contaminated urban soils with a sustainable industrial waste product.

There exists an entire subdiscipline of bioremediation called mycoremediation (myco=fungus) that focuses on using fungi to degrade toxic pollutants. Fungi make up a kingdom of life distinct from both plants and animals that includes yeasts, molds, and what are commonly referred to as mushrooms (Stamets, 2005). Growing on dead woody debris, mushrooms perform the essential ecological function of breaking down dead trees and making their nutrients available to other organisms. As nature's decomposers, fungi produce powerful enzymes capable of degrading both lignin and cellulose, two biological compounds that make up wood. These enzymes are also known to be capable of degrading a wide range of organic pollutants, including ones that may be otherwise resistant to microbial degradation alone. One of the challenges of mycoremediation is getting fungi to grow expansively through non-sterile media, such as the soils of a contaminated lot. Furthermore, it's important to be extremely cautious about eating any mushroom that grew in contaminated soils because they are known to accumulate toxic metals (Kirchner, 1998).

One method for getting around this problem is to grow mushrooms on a "clean" substrate separately and only use the leftover material for remediation. Mushrooms, particularly oyster mushrooms (*Pleurotus ostreatus*), are especially adept at growing on a wide range of materials including cardboard, straw, old coffee grounds, toilet paper, twine, human hair, etc. (Sánchez, 2010). A number of these are common urban waste products that can be acquired for free and would otherwise end up in the landfill. Coffee grounds are a favorite of mine as the work and energy of pasteurizing them (killing off competing fungi) is already done for free through the heat of brewing the coffee. Oyster mushrooms can then be grown on these materials, eaten and/or sold. Once the fungus has exhausted all the available nutrients in the substrate, it will no longer produce fruiting mushrooms. The enzymes it exuded, however, will still be present in the leftover substrate. This spent mushroom compost can then be spread across an area with organic pollutants, where the rains will wash the enzymes into the soil where they can facilitate the degradation of toxins (Gasecka, 2013); (Lau, 2003); (Mohammadi-Sichani, 2018); (Okerentugba, 2015); (Purnomo, 2010). This spent mushroom compost is also an excellent agricultural amendment, so at minimum soil conditions would be improved through its application.

Using this technique, it is possible to create an ecologically regenerative micro-industry:

(1) Organic wastes are collected and kept out of the landfill, preventing the creation of greenhouse gases.
(2) Edible mushrooms are grown and either eaten or sold, thereby providing both food and jobs to urban communities.
(3) By-products are used to regenerate the health of contaminated urban soils.

Another excellent reason to grow mushrooms in cities is the fact that that they do not require sunlight for photosynthesis. For this reason, they can be grown in marginal and shady environments that would be inappropriate for vegetable production such as basements, alleyways, shaded lots, the north sides of buildings, and obsolete industrial warehouses (Garnett, 2000). Warehouses are great places to grow mushrooms as they can be grown in hanging bags that go floor to ceiling (Figure 3.4). Waste products could then be spread on the soils surrounding the building and eventually be used for farming.

It is possible to also just collect the wastes from a commercial mushroom operation and use them for remediation. The Radix Center has done this using the wastes from Ecovative, a company based in Green Island, NY that produces a polystyrene replacement product made from fungal mycelium. After signing a non-disclosure form, Radix was permitted to dumpster dive at their factory and experiment with using their discards for remediation.

Instructions for simple cultivation of oyster mushrooms

What you will need:

- One plastic storage container with lid
- One 2-pound bag of oyster mushroom spawn
- Approximately five gallons of relatively fresh used coffee grounds (can often be acquired for free from coffee shops)

What to do:

Step 1: make several drainage holes in the bottom of the storage bin.

Step 2: mix coffee grounds and mushroom spawn together thoroughly in plastic storage bin, filling it to its top.

Step 2: cover bin and store at 60–80 degrees Fahrenheit for approximately three weeks. Add water periodically if moisture levels drop below 40 percent (equal to a wrung-out sponge).

Step 3: check periodically on growth of the fungal culture. When mushroom formation begins, ensure humidity levels remain at 80–100 percent (can be maintained by creating a "humidity tent" made from ventilated plastic bags).

Mealworms and polystyrene

The global production and improper disposal of plastics is an enormous environmental problem, with many plastics being resistant to degradation and contributing to the pollution of soils, rivers, lakes, and oceans. One particularly recalcitrant plastic is polystyrene. More commonly known by its proprietary name Styrofoam, polystyrene is a polymer commonly used for insulation, packaging, and food containers. It is produced and disposed of in large quantities exceeding more than 20 million tons a year (PlasticsEurope, 2016), is toxic, and is extremely resistant to degradation under normal environmental conditions.

Figure 3.4 Oyster Mushroom Cultivation

Source: Scott Kellogg

While plastics do have long environmental residence times, they will degrade under certain circumstances. The ability of insects to break down plastics has been known for some time but has only been studied as a problem of the plastics manufacturing industry and not as a potential solution to plastics pollution. Recent studies, however (Yang Y. e., 2015); (Yang S.-S. e., 2018), have demonstrated the ability of mealworms, a type of insect, to effectively degrade polystyrene. Could this be a possible bioremediation technique for addressing the problem of plastics?

Mealworms are the larval state of the darkling beetle (*Tenebrio molitor*) and are commonly raised as a feedstock for reptiles, fish, and birds (they are also perfectly edible for humans as well). They can be reared easily in small containers by being fed a diet of bread, oats, or bran and a slice of potato or carrot as a moisture source. As mentioned, a handful of recent studies have shown that mealworms amazingly have the capability to degrade polystyrene, turning it into carbon dioxide and a fecal pellet (Figure 3.5). The degradation is done largely by bacteria living in the gut of the mealworm, making this an interesting instance of insect–microbe bioremediation symbioses (Tang, 2017). While studies suggest that mealworms can live off polystyrene alone for a period, they do better when their diet is supplemented by something else, such as bran.

While this is certainly an interesting discovery, it's by no means a perfect solution to the problem of plastics. It will be important to proceed cautiously before advocating for building giant mealworm-polystyrene-digester-reactors. For one thing, the polystyrene is not completely broken down. In fact, a fair amount of it is ejected from the mealworm as a fecal pellet. This means that polystyrene-fed mealworm frass (the technical term for insect droppings) would be undesirable as a fertilizer or for composting, as it contains quite a bit of tiny pulverized polystyrene particles. Has the plastic problem just been made worse by breaking it down into millions of tiny pieces? Perhaps – but a lot of it has been effectively mineralized in the process. Another important question is whether there is any toxic residue from the process left inside the body of the mealworm. This is something to be particularly cautious about, especially before using mealworms as a food for livestock where toxins could biomagnify.

Figure 3.5 Mealworms at Work

Source: Scott Kellogg

While we primarily should be working to end the global production of non-compostable toxic plastics, perhaps these studies offer solutions for what could be done to remediate plastic pollutants that already exist using simple, decentralized, and affordable technologies. Studies suggest that other species of insects could also be capable of degrading other types of plastics, and that a second generation of plastic-fed bugs could be even more effective plastic eaters than their parents. At minimum, using mealworms to eat polystyrene can be done on a small and experimental scale and creates an interesting entry point into bioremediation education.

Instructions for setting up a small-scale mealworm polystyrene digester

The following are basic instructions for building a small mealworm farm that can be used to demonstrate the principles of biologically assisted degradation of synthetic polymers.

What you will need:

- One cup of mealworms (approximately 100 larvae). Readily available at pet stores as a reptile feed
- One plastic storage container with lid (size does not matter hugely)
- One package of oatmeal and a carrot
- One piece of polystyrene (plate, cup – preferably recovered from the waste stream)

What to do:

Step 1: place mealworms, oats, a slice of carrot (provides moisture for the larvae), and polystyrene together in the storage container and maintain at room temperature.

Step 2: observe the rate of degradation of polystyrene.

Advanced:

Separate mealworms into three separate containers having:

(1) Only oats
(2) Oats and polystyrene
(3) Only polystyrene

Observe how mixes of different feedstocks affect mealworm growth and rates of pupation.

Bioremediation of metals

Bioremediation of toxic metals is a little more difficult than organic pollutants. As metals are elements, they cannot be broken down into any simpler or safer components, at least not under conditions found in the earth's biosphere. As a result, options for dealing with metals with bioremediation are limited to organically binding them up, reducing their mobility, or attempting to pull them out of soils using a process called phytoextraction. I've already described the first two approaches in the section on adding compost in risk reducing strategies, which leaves phytoextraction, a technique that is problematic in several ways. In theory, phytoextraction works by causing metals in the soil to become mobilized (usually by adding either acids or bases for cationic or anionic metals, respectively). Certain plants, known as "hyper-accumulators" would then "extract" the metals out of the soils and store them in their tissues. The plants would then need to be harvested, roots and all, and landfilled. In this sense, the problem of metals contamination is not being "solved" but rather moved from one place to another (arguably it's possibly better to have contaminants in a sealed landfill than in the top few inches of garden soils).

Phytoextraction's overall effectiveness as a remediation strategy is controversial. While some plants may be able to pull metals out of the soil, it is usually achieved by adding some type of chelating agent to the soil, a process that causes serious disruption to soil life, therein possibly making the "cure worse than the disease" (although soil pH could theoretically later be corrected). Another issue is that there is a "diminishing returns" effect with phytoextraction: it seems that initially high amounts of metals can be pulled from soils in the first several crops of plants in a phytoextraction trial, but then the levels of extracted metals start to drop, making it harder to remove the last remaining metals. For this reason, phytoextraction seems capable of only partial site remediation.

For these reasons, phytoextraction seems to be falling out of favor with practitioners and is being replaced by simpler methods such as compost amendment. I believe, however, in taking a balanced look at it as a technology and considering it as one among several tools at the disposal of bioremediation activists.

Testing and monitoring – making visible the invisible

One of the most insidious aspects of environmental contamination is its relative invisibility – a site may appear to be perfectly healthy while it may in fact harbor a toxic secret. Likewise, a site that people may believe to be contaminated may in fact not be at all. This uncertainty and lack of knowing manifests as fear among a population, contributing to ecological alienation and an impediment to building urban ecosystem justice.

While some forms of pollution are easily observed (oil spills, sewage leaks, smoke stacks), many times pollution is undetectable by human senses – contaminants are often invisible, odorless, and tasteless. Without sophisticated monitoring equipment, their presence may go unnoticed until their toxic health impacts manifest (sometimes only in future generations). This not-knowing is compounded by resource constraints: careful environmental monitoring is an expensive process involving precision technologies and skilled operators (at least when carried out by professional engineering firms). Furthermore, there is no single test which can detect every conceivable contaminant, each must be screened for individually at an enormous cost. Communities lacking funds or (in the case of many Global

South nations) testing laboratories altogether will be limited in sensing the scope of contamination in their soils, water bodies, and air.

One aspect of appropriate and community-based bioremediation is the development of accessible sensing technologies that are better capable of making toxicity visible and letting us know if remediation strategies are effective or not.

Soil testing

The cost of testing for soil contaminants makes it prohibitively expensive to most people. Standard soil tests offered by cooperative extensions typically test only for soil macronutrients (NPK), lead (Pb), and occasionally other metals. Testing beyond these basic parameters is extremely costly, as each contaminant, both organic and heavy metal, must be tested for individually. While such comprehensive soil testing can be carried out by engineering firms through what are called phase II environmental assessments, the service typically costs several thousand dollars. Doing background research on the past uses of a site can help to narrow the range of possible contaminants and reduce overall expenses.

The high cost of soil analysis not only makes it difficult to get a precise reading on the existing extent of soil contamination on a site but also as to whether a remediation strategy is effective in reducing contaminant levels.

At the Radix Center, we host events called "soil shops" (Figure 3.6) where community members are invited to bring samples of soil from their backyards to be screened for lead at no cost using an XRF owned by the Department of Health. Doing this helps to reduce the barriers of cost and inconvenience related to sending off a soil sample to a laboratory for testing. Furthermore, the soil shop is conducted in a very pro-gardening spirit: if a person's soil comes back with high lead

Figure 3.6 Soil Shop at Radix

Source: Scott Kellogg

levels, they are not scared into not gardening but rather informed and encouraged to take precautions to ensure that they can garden safely.

Conclusion: toxisphere versus ecoremediation

There are numerous obstacles that stand in the way of the rapid implementation of urban ecosystem justice. Most of them, and perhaps the most difficult to overcome, are social, cultural, and political. Often, the technical solutions to socio-ecological problems already exist, we simply lack the political will or the cultural fortitude to put them into action. One problem, however, of a mostly technical nature that remains persistent and pernicious is that of toxicity in the urban environment. This toxicity, commonly the by-product of urban industrial processes but also resulting from "inadvertent" sources such as paint, auto parts, and gasoline, exists as a substantial barrier to forming regenerative and mutually symbiotic relationships with the urban ecosystem. Cutting across all media, the "toxisphere" (an imperfect term of my creation) pervades air, water, soils, the waste stream, and living things alike. In truth, the toxisphere represents more of an idea than any one contaminant from any single source – it is more the collective and synergistic impact, both physical and psychological, of many pollutant streams combined. Manifesting often as fear, uncertainty, and as genuine health malady, anxiety over the toxisphere is an impediment to wanting to eat food from urban soils, swim in its rivers, breathe its airs, or befriend its creatures (for example brownfields are defined as properties that are unutilized either due to existing contamination or merely the *perception* of one). Uncertainty is one of the most vexing aspects of the toxisphere – often there is missing, incomplete, or inaccurate information regarding the impacts of toxicity in the urban ecosystem. Much of it is invisible. Detecting it requires access to expensive and sophisticated equipment, along with a level of technical expertise possessed by only a few (frequently private engineering firms whose services are unaffordable to most). Governmental agencies tasked with monitoring pollution are frequently overburdened, under-resourced, and unable to detect and map the extent of every form of pollution. When they do, their focus is on major point sources of emissions or on former industrial sites with leaking storage tanks that pose immediate and acute threats to public health. Less attention is put on "diffuse" pollutants: lead paint in backyard soils, street dust laden with heavy metals, PCBs in fish tissues, exhaust particulates on the surfaces of vegetables, groundwater contaminants being uptaken into the roots of trees. Furthermore, there is a high level of uncertainty regarded the fate and transport of toxic materials – it's one thing to identify the presence of a pollutant in the soil, but whether it is uptake into plant tissues or causes a threat to human health is entirely a different question. Because of this overarching uncertainty, the default response of public health agencies seems to be to "err on the side of caution" when any doubt exists and to refrain from activities that might put a person in direct contact with potentially contaminated media.

While I don't question that this default position makes sense under the logic of the juridical system we organize around, I'm seeking to raise questions about the factors considered in the risk assessment calculus used for making these decisions.

For one thing, assessment should consider the numerous potential *benefits* of activities such as gardening – nutritious food, physical activity, being outdoors, social connections, and educational value. When considered, it creates a stronger incentive to look at soils more thoroughly, to figure out how they can be tested affordably, and if a moderate contamination problem exists, offers a range of mitigating or "safer soil" practices that can reduce the risk of exposure rather than discouraging any gardening activity at all. Can the same logic be applied to areas beyond soils? Water, atmospheres, non-human life? Developing these models is an essential component of an urban ecosystem justice strategy.

It is here that I will introduce the concept of "ecoremediation" as a curative framework for addressing the toxisphere. Going beyond the idea of community-based bioremediation, ecoremediation examines the question of ecological remediation through a complex, whole systems lens in a manner that conventional public health, engineering, and bioremediation approaches alone do not. Whereas the prefix "bio" implies biology (a highly reductionist/deterministic discipline), "eco" implies ecology, a discipline with a considerably more holistic outlook. Ecoremediation does not reject the methods and techniques of bioremediation. They are, on the contrary, integral tools to the ecoremediation process. However, while the latter seeks to use one single species to degrade a contaminant, the former counts upon the combined, synergistic, and emergent effects of the co-metabolic processes of a huge diversity of organisms representing multiple pathways from the tree of life. Furthermore, ecoremediation considers the remediating impacts of ecologically regenerative practices such as sustainable agriculture, biodiversity conservation, and water remediation.

Ecoremediation is additionally better able to account for issues of uncertainty, ignorance, and surprise than bioremediation. Often, particularly in low-income communities, the resources to conduct extensive contaminant data collection and monitoring simply don't exist. Even the affordable "DIY" testing equipment employed by citizen scientist groups often provides too little accuracy or precision to be of significant value. Ecoremediation looks to other markers of ecosystem health as indicators of remediation: species diversity (bioassays), plant germination success rates, soil color, texture, and odor. While these indicators may not always successfully reveal the presence and concentrations of contaminants, they offer communities some yardstick by which to measure environmental health.

Ecoremediation also looks at remediation on a scale larger than a single lot, property, or farm. This scope ranges as large as the global and can be applied as well to entire cities, villages, and neighborhoods. On a global scale, ecoremediation acknowledges that all the remediation cleanup efforts that citizens are engaged in are limited in their effectiveness if the worldwide production of toxic materials goes unchallenged. Before you clean up the mess, you need to turn off the tap. Furthermore, an ecoremediation approach insists that corporate polluters must be held accountable for cleanup and reparations to the environments and communities who have been harmed by their actions. It does, however, try to simultaneously empower those groups with tools for remediation that can be used in the interim. Moreover, there needs to be a shift/return to using materials that are nontoxic and are ultimately compostable, degrading to organic soil. In this regard, ecoremediation is commensurable with ideas such as the precautionary principle as applied to toxic manufacturing.

Ecoremediation considers the need for "whole neighborhood remediation". For instance, because of the aerial deposition of dust-borne contaminants, it may be insufficient to produce clean soil in a single backyard if lead-based paint is chipping off surrounding buildings and blowing in. Furthermore, toxic street dust, resulting from the wearing of brake pads and car parts, poses an ongoing threat to the ecological health of (particularly urban) communities. Over time, this can result in a net increase in toxicity in an urban ecosystem.

Certain communities, particularly in the Global South, are suffering from the effects of multiple toxic legacies, which in themselves produce unknown synergistic effects. These communities are in many instances unable to afford or provide themselves the technical knowledge, monitoring equipment, and laboratory analysis needed to know the extent of this contamination and whether their attempts at remediation are being effective or not (at least in producing a reduction of specific contaminants, as would be verified through lab analysis). Ecoremediation accepts this uncertainty, aware of the knowledge gaps that exist. Recognizing, however, that many people's very survival is dependent on agriculture, gathering, hunting, and fishing from the land and water, an ecoremediation approach places a high importance on taking action to clean soils and water, even if the results and outcomes are not fully known. (In this regard, ecoremediation is less apt to accept that an area has been rendered as an "ecological sacrifice zone" and will employ a continuum of techniques that will enable people to continue subsistence lifeways that result in the least harm.)

In summary, an ecoremediation approach to toxisphere remediation could be described as "do everything you can to promote ecological regeneration and assume some benefit will result". This contrasts with the conventional approach which could be characterized as "we don't know the full extent of the contamination or necessarily what to do about it, therefor it's better to do nothing and just keep people away from there".

Over time, and through greater sharing of techniques and information, citizen bioremediation/ecoremediation could play a significant role in promoting the agendas of both the environmental justice and urban agriculture communities. Through forming mutually beneficial partnerships between humans, plants, fungi, and microbes in the process of remediation, the critical interdependencies necessary for socio-ecological health can be reestablished. It is my intention that this chapter could help in provoking a broader conversation about the challenges, strategies, and opportunities for achieving this.

Works cited

Alloway, B. J. (2004). Contamination of soils in domestic gardens and allotments: A brief overview. *Land Contamination and Reclamation*, *12*(3), 179–187.

Antisari, L. V. (2015). Heavy metal accumulation in vegetables grown in urban gardens. *Agronomy for Sustainable Development*, 35(3), 1139–1147.

Atkins, P. J. (2012). *Animal cities: Beastly urban histories*. Farnham: Ashgate Publishing, Ltd.

Atlas, S. (2015). *Soil Atlas 2015*. Retrieved from https://gsf.globalsoilweek.org/soilatlas-2015.

Bacon, A. R. (2011). Human – soil relations are changing rapidly: Proposals from SSSA's cross-divisional soil change working group. *Soil Science Society of America Journal*, 75(6), 2079–2084.

Bernes, C. (1998). *Persistent organic pollutants*. Stockholm: Swedish Environmental Protection Agency.

Birkhofer, K. e. (2008). Long-term organic farming fosters below and aboveground biota: Implications for soil quality, biological control and productivity. *Soil Biology and Biochemistry*, 40(9), 2297–2308.

Bledsoe, C. S. (2014). Beyond mutualism: Complex mycorrhizal interactions. In *Progress in botany* (pp. 311–334). Berlin Heidelberg: Springer.

Cao, H. e. (2010). Heavy metals in rice and garden vegetables and their potential health risks to inhabitants in the vicinity of an industrial zone in Jiangsu, China. *Journal of Environmental Sciences*, 22, 1792–1799.

Chemlal, R. T. (2012). Microbiological aspects study of bioremediation of diesel-contaminated soils by biopile technique. *International Biodeterioration & Biodegradation*, 75, 201–206.

Chen, G. (2009). *Alleviation of soil compaction by Brassica cover crops*. Ph.D. Dissertation, University of Maryland.

Chiang, F. H. (2013, Spring). Vermitea remediation of hydrocarbon contaminated soil. *Vermitea and Bioremediation*. Retrieved from https://nature.berkeley.edu/classes/es196/projects/2013final/ChiangF_2013.pdf.

Cookson Jr, J. T. (1995). *Bioremediation engineering: Design and application*. New York: McGraw-Hill, Inc.

Daniel, R. (2004). The soil metagenome: A rich resource for the discovery of novel natural products. *Current Opinion in Biotechnology*, 15(3), 199–204.

Darwish, L. (2013). *Earth repair: A grassroots guide to healing toxic and damaged landscapes*. Gabriola, BC: New Society Publishers.

Diamond, J. (2005). *Collapse: How societies choose to fail or succeed*. New York: Penguin.

DiSalvo, C. (2014). 17 The Growbot Garden Project as DIY speculation through design. *DIY Citizenship: Critical Making and Social Media*, 237.

Drobník, J. (1999). Genetically modified organisms (GMO) in bioremediation and legislation. *International Biodeterioration & Biodegradation*, 44(1), 3–6.

Egendorf, S. P. (2018). Constructed soils for mitigating lead (Pb) exposure and promoting urban community gardening: The New York City Clean Soil Bank pilot study. *Landscape and Urban Planning*, 175, 184–194.

Eglash, R. a. (2014). Basins of attraction for generative justice. In *Chaos theory in politics* (pp. 75–88). Dordrecht, the Netherlands: Springer.

Funtowicz, S. O. (1999). Science for the post-normal age. *Futures*, 25(7), 739–755.

Garnett, T. (2000). Urban agriculture in London: Rethinking our food economy. In *Growing cities, growing food* (pp. 477–500). Feldafing, Germany: German Foundation for International Development.

Gasecka, M. e. (2013). The efficient degradation of selected PAHs in soil with a substrate refuse from *Pleurotus ostreatus* cultivation. *Fresenius Environmental Bulletin*, 22(9A), 2651–2658.

Gittleman, M. e. (2017). Estimating stormwater runoff for community gardens in New York City. *Urban Ecosystems*, 20(1), 129–139.

Haller, H. e. (2016). Microbial transport of aerated compost tea organisms in clay loam and sandy loam: A soil column study. *International Biodeterioration & Biodegradation*, *106*, 10–15.

Hess, D. J. (2016). *Undone science: Social movements, mobilized publics, and industrial transitions*. Cambridge, MA: Massachusetts Institute of Technology Press.

Hillel, D. (1992). *Out of the earth: Civilization and the life of the soil*. Berkeley, CA: University of California Press.

Hou, D. a.-T. (2014). Sustainability: A new imperative in contaminated land remediation. *Environmental Science & Policy*, *39*, 25–34.

Hough, R. L. (2004). Assessing potential risk of heavy metal exposure from consumption of home-produced vegetables by urban populations. *Environmental Health Perspectives*, *11*.

Jonsson, A., & Haller, H. (2014). Sustainability aspects of in-situ bioremediation of polluted soil in developing countries and remote regions. *Environmental Risk Assessment of Soil Contamination*, 57–86.

Jouquet, P. e. (2006). Soil invertebrates as ecosystem engineers: Intended and accidental effects on soil and feedback loops. *Applied Soil Ecology*, *32*(2), 153–164.

Kellogg, S. (2016). Community based bioremediation: Grassroots responses to urban soil contamination. *Teknokultura*, *13*(2), 491–510.

Kirchner, G. a. (1998). Accumulation of 210Pb, 226Ra and radioactive cesium by fungi. *Science of the Total Environment*, *222*(1–2), 63–70.

Kubiëna, W. L. (1970). *Micromorphological features of soil geography*. No. 631.42 K8.

Lau, K. L. (2003). Use of spent mushroom compost to bioremediate PAH-contaminated samples. *Chemosphere*, *52*(9), 1539–1546.

Lidsky, T. I. (2003). Lead neurotoxicity in children: Basic mechanisms and clinical correlates. *Brain*, *126*(1), 5–19.

Liittschwager, D. (2012). *A world in one cubic foot: Portraits of biodiversity*. Chicago, IL: University of Chicago Press.

Margulis, L. (2009). Genome acquisition in horizontal gene transfer: Symbiogenesis and macromolecular sequence analysis. In *Horizontal gene transfer* (pp. 181–191). New York: Humana Press.

McClintock, N. (2010). Why farm the city? Theorizing urban agriculture through a lens of metabolic rift. *Cambridge Journal of Regions, Economy and Society*, *3*(2), 191–207

Menalled, F. e. (2009). *From conventional to organic cropping: What to expect during the transition years*. Bozeman, MT: Montana State University Extension Service MontGuide MT200901AG.

Mielke, H. W. (2012). The urban rise and fall of air lead (Pb) and the latent surge and retreat of societal violence. *Environment International*, *43*, 48–55.

Miliutenko, S. A. (2013). Opportunities for environmentally improved asphalt recycling: The example of Sweden. *Journal of Cleaner Production*, *43*, 156–165.

Mohammadi-Sichani, M. e. (2018). Ability of *Agaricus bisporus*, *Pleurotus ostreatus* and *Ganoderma lucidum* compost in biodegradation of petroleum hydrocarbon-contaminated soil. *International Journal of Environmental Science and Technology*, 1–8.

Monosson, E. (2012). *Evolution in a toxic world: How life responds to chemical threats*. Washington, DC: Island Press.

Ohsowski, B. M. (2012). The potential of soil amendments for restoring severely disturbed grasslands. *Applied Soil Ecology*, *60*, 77–83.

Okerentugba, P. e. (2015). Spent mushroom compost for bioremediation of petroleum hydrocarbon polluted soil: A review. *Journal of Environmental Science and Toxicology*, *4*(1), 1–7.

O'Riordan, T. (2013). *Interpreting the precautionary principle*. London: Routledge.

Perelo, L. W. (2010). In situ and bioremediation of organic pollutants in aquatic sediments. *Journal of Hazardous Materials*, *177*(1–3), 81–89.

Pirkle, J. L. (1998). Exposure of the US population to lead, 1991–1994. *Environmental Health Perspectives*, *106*(11), 745.

PlasticsEurope. (2016). *Plastics-the facts*. Retrieved October 20, 2016, from http:www.plasticseurope.org/plastics–the-facts-2016-15787.aspx?FollD

Prasad, L. R. (2000). Contamination of allotment soil with lead: Managing potential risks to health. *Journal of Public Health*, *22*(4), 525–530.

Puig de la Bellacasa, M. (2015). Making time for soil: Technoscientific futurity and the pace of care. *Social Studies of Science*, *45*(5), 691–716.

Purnomo, A. S. (2010). Application of mushroom waste medium from *Pleurotus ostreatus* for bioremediation of DDT-contaminated soil. *International Biodeterioration & Biodegradation*, *64*(5), 397–402.

León Rejón, L., García Trigueros, B., Rosales Reyes, M., Soto Paniza, C., Cámara Pérez-Vela, C., Cifuentes, C., . . . & Delgado-Moreno, L. (2013). Use of vermitea for removing pesticides from soil. *High School Students for Agricultural Science Research*. Volume III is licensed under a Creative Commons Reconocimiento-SinObraDerivada 3.0 Unported License: 69.

Renforth, P. e. (2011). Designing a carbon capture function into urban soils. *Proceedings of the ICE-Urban Design and Planning*, *164*, 121–128.

Richter, D. e. (2015). Soil in the Anthropocene. In *IOP conference series: Earth and environmental science* (p. 25). Bristol: IOP Publishing.

Rondon, M. R. (1999). The earth's bounty: Assessing and accessing soil microbial diversity. *Trends in Biotechnology*, *17*(10), 403–409.

Sánchez, C. (2010). Cultivation of *Pleurotus ostreatus* and other edible mushrooms. *Applied Microbiology and Biotechnology*, *85*(5), 1321–1337.

Santos-Lozada, A. R. (2018). In Puerto Rico, counting deaths and making deaths count. *Health Affairs*, *37*(4), 520–522.

Sargent, J. D. (1994). The role of nutrition in the prevention of lead poisoning in children. *Pediatric Annals*, *23*(11), 636–642.

Scalenghe, R. a. (2009). The Anthropogenic sealing of soils in urban areas. *Landscape and Urban Planning*, *90*(1), 1–10.

Séré, G. e. (2010). Early pedogenic evolution of constructed Technosols. *Journal of Soils and Sediments*, *10*(7), 1246–1254.

Shiva, V. (2008). *Soil not oil: Environmental justice in a time of climate crisis*. Cambridge, MA: South End Press.

Smith, N. J. (1980). Anthrosols and human carrying capacity in Amazonia. *Annals of the Association of American Geographers*, *70*(4), 553–566.

Stamets, P. (2005). *Mycelium running: How mushrooms can help save the world*. New York: Random House Digital, Inc.

Stoessel, F. e. (2018). Assessing the environmental impacts of soil compaction in Life Cycle Assessment. *Science of the Total Environment*, *630*, 913–921.

Storelli, M. M. (2008). Potential human health risks from metals (Hg, Cd, and Pb) and polychlorinated biphenyls (PCBs) via seafood consumption: Estimation of target hazard quotients (THQs) and toxic equivalents (TEQs). *Food and Chemical Toxicology*, *46*(8), 2.

Tang, Z.-L. T.-A.-H. (2017). The study of the microbes degraded polystyrene. *Advances in Technology Innovation*, *2*(1), 13–17.

Targulian, V. O. (1978). Structural and functional approaches to the soil: Soil-memory and soil-moment. In *Mathematical modeling in ecology* (pp. 17–33). Moscow: Nauka.

Tate, R. L. (1995). *Soil microbiology*. Hoboken, NJ: John Wiley and Sons.

Tilman, D. (1998). The greening of the green revolution. *Nature*, 211.

USEPA. (2009). How does your garden grow? *Brownfields Redevelopment and Local Agriculture*. Retrieved April 6, 2011, from www.epa.gov/brownfields/success/local_ag.pdf.

Van Der Heijden, M. G. (2008). The unseen majority: Soil microbes as drivers of plant diversity and productivity in terrestrial ecosystems. *Ecology Letters*, *11*(3), 296–310.

Vasenev, V. a. (2018). Urban soils as hot spots of anthropogenic carbon accumulation: Review of stocks, mechanisms and driving factors. *Land Degradation & Development*, *29*(6), 1607–1622.

Vidali, M. (2001). Bioremediation: An overview. *Pure and Applied Chemistry*, *73*(7), 1163–1172.

Wakefield, S. Y. (2007). Growing urban health: Community gardening in South-East Toronto. *Health Promotion International*, *22*(2), 92–101.

Walker, L. R. (2012). *The biology of disturbed habitats*. Oxford: Oxford University Press.

Washbourne, C.-L. P. (2012). Investigating carbonate formation in urban soils as a method for capture and storage of atmospheric carbon. *Science of the Total Environment*, *431*, 166–175.

Yang, S.-S. e. (2018). Biodegradation of polystyrene wastes in yellow mealworms (larvae of Tenebrio molitor Linnaeus): Factors affecting biodegradation rates and the ability of polystyrene-fed larvae to complete their life cycle. *Chemosphere*, *191*.

Yang, Y. e. (2015). Biodegradation and mineralization of polystyrene by plastic-eating mealworms: Part 1, Chemical and physical characterization and isotopic tests. *Environmental Science & Technology*, *49*(20), 12080–12086.

4 Urban waters justice

DIY river remediation in the aquatic commons

Introduction

Albany, NY, Summer 2018 – The news cameras closely followed the group of youth as they carried the floating island down to the banks of the Hudson River near the edge of the heavily industrialized Port of Albany. Resembling a mass of plastic tubing intertwined with swamp plants, the island was placed in the water and dragged out with kayaks and canoes to be strategically placed near a massive three-foot diameter sewage discharge pipe. Already in the river was a solar panel-equipped barrel raft tied to several other smaller floating islands. Once all the pieces were coupled together, makeshift anchors created from concrete blocks were dropped on either side, holding the island in a southward-facing orientation. Now fully energized, a solar-powered pump hummed away, blowing beautiful tiny bubbles into the roots of the plants from the murky depths below. The artful-yet-functional techno-ecological hybrid assemblage known as the "artificial floating island" was now complete. As the crowd of onlookers applauded and cheered, I could feel the sense of satisfaction and pride that the youth took in their accomplishment (Figure 4.1).

With hope, this creation would help to degrade sewage and stormwater pollutants that impaired the health of the river. At minimum, it was a participatory project that fostered a sense of care, love, and responsibility between local youth and the well-being of the Hudson, who despite having grown near its shores, few had ever visited. Focusing on the floating island provided a way to shine a light on persistent infrastructural problems such as combined sewage overflows, illustrating the connections between urban ecosystem health and concerns of social justice, access, and equity and weaving in broader conversations about issues of exclusion and alienation in the urban ecosystem. Who owns the river? How do you access it? Who is responsible for it being polluted? What can we do, as ordinary people, to heal it and reestablish an aquatic commons? In order to better understand the relationships between people, water, and waste in cities, it is important to first understand their entangled histories and how they've been broken apart over time and what steps are necessary for bringing them back together.

DOI: 10.4324/9780367858711-5

Figure 4.1 Launch of the Island

Source: Scott Kellogg

Water in cities

Water is itself both alive and essential for all life. A complex ecology of elements, nutrients, dissolved gases, plants, animals, and microbes, water that is vital yields great abundance while water that is poisoned and lifeless provides very little. Not only are human bodies incapable of surviving more than a few days without water, aquatic life provides people with an invaluable source of nutrition (Logsdon, 1978). Estuaries, the environments where seawater and freshwater mix, are among the most biologically productive ecosystems on the planet (Beatley, 2014). Fish, shellfish, crustaceans, and the animals that feed off them are all abundant at these sun-soaked and nutrient-rich confluences of rivers and seas. Living in a buoyant environment that reduces the stresses of gravity, aquatic organisms can direct more of their energy into growth, making them highly efficient converters of protein to body mass (McLarney, 1984). Agriculture of any sort depends upon the reliability of water, provided through either rainfall or irrigation. As such, permanent settlements are rarely found far from a dependable water supply (Flannery, 2002). For drinking, farming, fishing, washing, transportation, trade, and industry, it's no wonder that most cities are built on the edge of a river, lake, or an ocean, providing essential needs for human sustenance.

Over time, however, on account of the growth of cities, industrialism, and auto infrastructure, urban waterways have become increasingly abused and neglected, treated like open sewers to carry away all manner of urban waste products. Toxins leached into waterways have poisoned the organisms living in them, making them unfit for fishing in, let alone swimming or drinking. Access to waterways has been cut off in many cities, with highways, fences, and barricades denying citizens a direct relationship with the water bodies that flow through their communities. During the 1950s and 1960s, numerous interstate highways were constructed that ran parallel to rivers. The highways, designed to promote rapid and efficient commuting for suburban dwellers and their white-collar urban office jobs, effectively cut off urban residents access to waterways. They also had the effect of dividing neighborhoods, cutting low-income residents off from essential city services, promoting urban sprawl, and producing dangerous levels of air pollution concentrated in poor neighborhoods (Noonan, 2005). As corporate interests have further enclosed the aquatic commons, a narrative has been simultaneously spun that urban waters are hopelessly contaminated and beyond salvation. Why fight to protect something if it is already beyond repair and alien to you?

Take the Hudson River as an example. Once a thriving tidal estuary with abundant aquatic life that sustained people over thousands of years, in Albany there are now advisories against eating most fish species out of the river, a consequence of PCBs (poly chlorinated biphenyls) leaching from industrial sites and building up in the food chain (Sloan, 1988). To make matters worse, the construction of Interstate Highway 787 in the mid-twentieth century has all but cut off residents' access to the riverside, a tragic decision made to facilitate suburban commuting at the expense of people's connection to the river. The inaccessibility and poisoning of waterways have had a disproportionate impact on lower-income urban residents and communities of color, who frequently rely on fish caught from them for sustenance and survival (Wattigney, 2017).

While the current state of urban waterways is bleak, there are reasons for hope. In the decades following the enactment of the Clean Water Act and the consequent end of unrestricted discharge of industrial pollutants, the health of rivers has rebounded considerably, with many once extirpated aquatic species now returning (Porter, 2015). Much work remains to be done, though, as many non-source point and diffuse pollution sources still threaten waters. Furthermore, while some urban communities have embraced their waterfronts by removing barriers to their access or through stream daylighting, building exclusive beaches and boathouses, these developments have largely benefited wealthier residents. The premium shoreline real estate that is created results in the gentrification of waterfront neighborhoods (Checker, 2018). In low-income communities, relatively little has been done to reconnect people to the water or to remediate the sources of toxicity that render them unfit (Curran, 2018).

For all these stated reasons, it is critically important to reconnect urban residents with waterways, teaching them about their ecology, history, and social significance while emphasizing matters of equity and inclusion. Developing this basic familiarity is an essential first step for regenerating the ecological health of urban waters and for pushing back on the idea that they are beyond hope. By addressing the systemic sources of waterway deterioration, and through citizen-directed initiatives guided by concern and love for the well-being of water, a relationship of reciprocity between humans and water in the city may be fostered. After enough time and effort, it may be possible for all residents to share in the joy and bounty of swimming, fishing, and cocreation with urban waters.

Stormwater, the CSO, and threats to the health of urban waters

A major water quality issue impacting urban watershed health is stormwater runoff, which occurs as a result of the ubiquity of impervious cover in urban areas in the form of roadways, parking lots, and rooftops. During a rainfall event, water that would normally infiltrate the soil instead lands on pavement and is diverted into storm drains, carrying with it a mixture of gasoline, pet feces, floating debris, lawn fertilizers, and pesticides (Tsihrintzis, 1997). This toxic mixture is then discharged into local waterways, where it severely impacts the health of aquatic ecosystems. In older cities with combined sewer systems, this problem is compounded when a heavy rainstorm triggers a combined sewage overflow (CSO), resulting in the discharge of unfiltered raw sewage and industrial wastes into waterways. The influx of sewage nutrients can cause an algae bloom and a subsequent depletion of oxygen, resulting in a eutrophic dead zone where no fish can survive. Furthermore, shellfish living in the spill zones of sewage overflows can trap pathogenic fecal organisms in their tissues, causing them to be inedible (Borchardt, 1997). As urban sprawl advances, this problem will only worsen. It will be further compounded by heavy rainfall events, or "rain bombs", that are predicted to occur with greater frequency as climate change progresses, a consequence of warms air's increased capacity to store moisture (Willems, 2012).

The framing of urban watershed quality issues through the context of stormwater runoff and combined sewage overflows is important as it gives a relatable lens for studying the problem and numerous opportunities for redress that result in mutually beneficial human/watershed relationships (Figure 4.2). Furthermore, there are significant environmental justice dimensions to the issue of CSOs, with it having been demonstrated that CSOs disproportionately impact low-income and African American communities in certain cities (Bullard, 1999). For this reason, CSOs are used as an entry point into a broader conversation about urban waterway justice.

The problem of stormwater runoff and sewage overflows can seem enormous and daunting. Small decentralized actions, however, such as building floating islands and rainwater collection systems can give people the feeling that they can make a change, even if they are just small first steps. Scaled up, these systems can make a significant impact on local water quality and can foster a greater sense of reciprocal connection and concern with the well-being of the river.

Figure 4.2 Warning Sign
Source: Scott Kellogg

Many of the citizen-centered approaches for addressing urban stormwater runoff involve the removal of impervious covers or the detainment and slow release of water. Both strategies can be accomplished in ways that are low-cost, involve minimal technological investment, and are decentralized, all and all making them more accessible to the average urban resident (Frazer, 2005). Furthermore, their implementation will be further incentivized by the added benefit of the development of ecologically productive and beautified spaces for human use and enjoyment.

As mentioned, impervious covers significantly impact the health of urban watersheds. The simple act of de-paving or tearing up of asphalt and the subsequent de-compaction of exposed soils can greatly offset the amount of stormwater directed into drains. By giving rainwater a chance to soak into the earth, it is retained, filtered, and slowly released back into waterways over a period of days, thereby dramatically reducing the "flashiness", or episodic pulsing and tendency to produce flash flood events, of the urban watershed (Moran, 2010). The de-paved area can then be made into a garden, park, or other green space, performing both valuable ecosystem services while beautifying the landscape. Youth in an ecojustice curriculum will be empowered by the relatively simple act of removing asphalt with pickaxes and shovels and will tangibly appreciate both the personal and ecological benefits of their labor. Similar results to de-paving can be produced through the construction of vegetated roofs and rooftop gardens, both of which can detain rainwater and create educational and food producing spaces for youth (Danks, 2010).

It should be mentioned that simply removing impervious pavements may be insufficient for offsetting stormwater runoff, as the soils beneath have often been severely compacted by the weight of vehicles driving across them over years. It may be necessary to de-compact as well as de-pave in order to increase these soils' absorptive capacity. While this can be done to some extent through physical and mechanical means (pickaxes, broad forks), it is intensely strenuous and likely impractical in a large area. It may be possible to rely on some plant species to perform the work of de-compaction, their roots being capable of breaking up hardpan soils. This technique requires far fewer labor inputs and will also do much to improve the general health of soils (Chen, 2010).

Improving physical access to waterways is a more difficult issue, as short of advocating for the removal or alteration of interstate highways, little can be done to improve citizens' access to waterways. The problem remains significant, however, as in many instances youth who have grown up near a waterway have never actually had any contact with it. While the construction of pedestrian bridges spanning highways is certainly desirable, the financial and political barriers in the way of doing so put this strategy beyond the reach of ordinary citizens. There are simpler strategies, though, that can be taken to help get around, if not remove, the barriers that block access to waterways. Simply organizing and providing transportation for outings to waterways can do a great deal to foster a respect, relationship, and concern for the well-being of waterways among youths. Boating and kayaking provide safe opportunities for the exploration of waterways, and a chance to be in more intimate contact with water. When youth become engaged in the environmental monitoring of waterways, they are enabled firsthand to see what impacts human activity can have on the health of aquatic environments and gain valuable skills in taking samples and undertaking data analysis. Additionally, the practices of culturing oysters, building floating restorers, or attending fishing clinics help to develop a sense of deeper care and affinity for the life in waterways and a sense of interwoven fate and interdependence.

Standout organizations involved in doing this type of work include Rocking the Boat, a New York City-based organization that maintains a boat house on the banks of the Bronx River. The group teaches high school age youth the basics of boating and navigation, environmental monitoring, and oyster reef restoration (Krasny, 2012). Another example is the Lower East Side Ecology Center, another New York City organization that sponsors fishing clinics that teach youth the basics of catch and release fishing, as well as fish identification and safety (LESEC, 2015).

Artificial floating islands – DIY river remediation

Artificial floating islands (AFIs), or floating wetland beds, are a technology that is used in ecological engineering practices to remediate eutrophied water bodies. While their design varies considerably, they essentially consist of a constructed raft of buoyant materials that suspend emergent wetland vegetation. The roots of the water plants are then permitted to grow downwards into the water column where they provide attachment sites for microbial biofilms, supported by the oxygen provided through the plant's vascular system. The attached microorganisms can remediate a number of waterborne pollutants, while the plants themselves uptake nutrients that are transformed into a harvestable biomass (Cao, 2010); (Zhao, 2012); (Guoyan, 2005); (Cao & Zhang, 2014). AFIs are modeled after natural floating islands, a phenomenon that occurs when a piece of a lake's shoreline is torn loose in a storm and remains afloat. Over time, the roots of plants will grow down in the water and help to purify it as it drifts about. As plant matter decomposes, trapped gases add to the island's buoyancy. Wildlife will also use the island as habitat, while young fish can use its roots for protection. By taking inspiration from nature, AFIs provide many of the same ecosystem benefits of natural floating islands.

The potential to involve youths and citizens in the construction of floating islands is great, as their essential components can be either salvaged from waste streams or collected from outdoor environments. Recycled plastic bottles stuffed into a plastic mesh tube create a floating structure. Alternatively, lengths of bamboo can be rafted together to produce a similar function. In either case, the construction is participatory and fun and can be artistically embellished. Upon completion, the island can be anchored near a sewage outfall where it will assist in the degradation of sewage and stormwater pollutants. If deployed early in the summer, youth will have the opportunity to observe the island growing and changing over time. Growing food on such an island is not recommended as plant roots would be in contact with polluted waters and could potentially uptake chemical and biological contaminants. At the end of the growth season, the island can be brought ashore and the biomass composted, effectively removing mobile nutrients from the water column. In this regard, floating islands provide an excellent way to build relations between urban youth and waterways, combine scientific and artistic properties, and foster concern and activism for the health of water.

Mechanical aeration is an optional feature for AFIs. Plants alone can deliver oxygen to the bacteria living on their roots, but by diffusing air into the roots of plants with an air compressor, the ability of the supported microorganisms to degrade pollutants is greatly enhanced.

The effectiveness of AFIs depends on the ratio of plant–root mass to the water they are in – a small three-foot diameter island is likely to have a minimal impact on a body of water as large as a river. AFI technology is highly modular and scalable – numerous smaller islands can always be rafted together later to have a larger impact on water quality. While a small island may not have a huge effect on the health of a large water body, its educational and symbolic value can still be enormous.

A wide range of aquatic plants may be used in an AFI. Native species with rapid growth rates, high rooting depth, and a large biomass will do best (Figure 4.3). Noninvasive plants should not be used so as not to accidently release them into a waterway. It is recommended that the maximum diversity of plant species be selected for in use of a floating island – greater diversity will initially improve the island's chance of success, as there is a better probability that at a least some of the species will survive and prosper.

Figure 4.3 Close-up of Island Roots

Source: Scott Kellogg

Design instructions

Included here are instructions for building an "entry level" floating island (Figure 4.4). Note that it does not have any component of active aeration – having aeration will certainly improve an island's performance but is not necessary. This "passive" design will rely upon only on the action of bacteria living attached to the roots of the plants.

What you will need:

- One 50′ roll of 1″ diameter polypropylene irrigation tubing
- One 10′ roll of 4′ high plastic chicken fencing
- A 1″ barbed poly-pipe connector
- 1.5″ hose clamps
- Assortment of zip-ties
- Water plants

What to do:

Step 1: coil the irrigation tubing into three or more loops of desired diameter (3 feet across is a good starter size). Secure loops with zip-ties.

Step 2: join the open ends of the tubing with the barbed connector and tighten with the hose clamps (this will prevent the tube from filling with water and losing its buoyancy).

Figure 4.4 Entry-level Island

Source: Scott Kellogg

Step 3: roll out plastic chicken fencing and place the looped tubing on top. Attach fencing to loops with zip-ties and cut off fencing that goes beyond the loops so that the entire form is circular.

Step 4: attach plants. Remove any rocks or soil from plant roots and attach to roots to plastic chicken fencing with zip-ties so that plants are standing upright. It may be initially necessary to use a supportive trellis or lightweight poles to ensure that plants remain upright.

Step 5: deploy. At this stage, there is a choice of either putting the island out into the water or letting it "incubate" in a protected container.

Incubation: Incubation gives a chance for the plants on the island to grow and for the microbial community to mature before being cast out into a river or pond. This can be done simply by placing it in a sunny spot in a kiddie pool filled with enough water to keep the roots wet. Once the plants have had time to adjust to transplant shock, the island can be put out.

Deployment: Here are some considerations for the location of the island's deployment:

One, make sure it doesn't float away. An anchoring system as simple as a rope tied to a concrete block can work. Be sure that it remains in an area that receives abundant sunlight, as most aquatic plant species require full sunlight. Two, make

Figure 4.5 On the Shores of the Hudson

Source: Scott Kellogg

sure it's not in a place where boats could collide with it. If there is a risk of this, consider flagging the island or placing solar LEDs on it to increase its visibility. Third, place it where it will be most beneficial, such as the near vicinity of a sewage or drainage pipe. Be mindful, however, of the high velocities of water coming out of these pipes that could cause islands to lose their anchorage and get washed away.

Floating islands can be launched from both boats and from the shoreline (Figure 4.5). A boat (even a canoe or kayak) gives the advantage of being able to drag and drop the island in any location. Islands can also be pushed off from shore. In this case, a relatively long anchor line that could be thrown out into the water from shore would be required.

AFI version 2

The following are the details of a larger-scale island constructed by the Radix Center that is described in the introductory paragraph.

Radix's AFI comprised simple floating island cells that were enhanced by an active aeration component. This consisted of a solar-powered air pump mounted on a wooden dock kept afloat by fifty-five-gallon barrels. The solar panel that powered the pump was also attached to the wooden dock. Space remained on the surface of the dock to be used for a person to conduct maintenance on either the island or the pump. Tubing leading away from the pump goes to two large air diffusers that were suspended beneath each of the largest wetland cells. This way, bubbles would rise into the root zones of the plants to assist them with contaminant degradation. Lastly, the island contained a GPS tracking device that could be used to recover the island should it ever be lost.

The design, construction, and implementation of Radix's AFI were carried out in conjunction with the "Ecojustice Summer" program, a five-week summer youth employment immersion that combines community gardening with sustainability justice education and outdoor adventure. By including the youth, not only did the island function as a tangible model of participatory, problem-based, and experiential learning, it also spurred interest in the history and ecology of the river itself. In this sense it operated as an effective means for challenging ecological alienation among the youth, many of whom, despite having grown up in Albany, had never actually stood on the banks of the Hudson, let alone gone out into it on a boat.

The islands, built cooperatively by the Ecojustice youth, were "incubated" in stock tank ponds at the Radix Center for several weeks before being deployed. This not only gave the chance for both plants and their attached microbes to mature but also allowed the youth to develop an intimate familiarity with the system, seeing it as a process which changes over time. By "inoculating" the islands with sediment samples (Figure 4.6) from nearby ponds and rivers, their microbial communities were given a head start before being deployed in the river.

Figure 4.6 Incubating the Islands

Source: Scott Kellogg

Rainwater harvesting – turning off the tap

While floating islands provide a fun and educational tool for remediating urban waterways and drawing attention to urban water quality issues, they are limited to containing and cleaning waters only after they have been polluted. A comprehensive approach to stormwater mitigation would strive to "turn off the tap" of stormwater pollutants in addition to "mopping up the mess". One other strategy for CSO prevention that addresses the issue of stormwater runoff at its source is rainwater harvesting. By collecting rainwater that falls on rooftops in a barrel or other container, it is possible to use it for irrigation purposes over time. By slowing the water's release, urban flashiness is reduced. In this way, the watershed benefits through reduced stormwater runoff and the individual benefits by gaining a free source of non-chlorinated water. Youth engaging in rainwater collection have the opportunity to learn about the planet's hydrological cycle in a manner that is relatable and tangible and can further understand the benefits of using non-chlorinated water on the health of microbial life in gardens (chlorine, a microbicide, works against organic gardening's goal to culture healthy and diverse microbial populations in soil) (Ingham, 2000).

Broader scale connections can be made by having relevant conversations about the causes of acid rain and the global trend of water privatization. Furthermore, conversations about the relative purity of rainwater can be used to frame issues of urban groundwater contamination, a condition that prevents the construction of wells in both urban and rural environments (the planet's hydrological cycle is a distillation process, many of the toxins found on surface and groundwater will not be present in rain) (Sedlak, 2016). In this way, a physical object such as a rain barrel can be used to frame multi-scalar conversations, making linkages between processes occurring on an applied community-based level with those taking place on a regional and global scale (Lundy, 2011).

Rainwater that presently rushes off of impervious rooftops could be detained through rainwater collection systems and allowed to slowly enter soils, decreasing the likelihood of triggering a CSO. Such a system could be as simple as a fifty-five-gallon barrel placed beneath the downspout of a gutter on a house. If the water is not to be used for irrigation, a small hole can be made in the bottom of the barrel that will allow the water to drain out slowly, hugely reducing the amount of rainwater entering a sewer in a heavy rain event.

Many "pocket gardens" built in small vacant lots in urban environments do not have water on-site. This can create serious problems during dry periods when it would be necessary to carry water to the gardens. At eight pounds a gallon, water is prohibitively heavy for many to carry. This problem could be solved by paying for a connection to a city water line, however costing as much as four thousand dollars, well beyond the budget of most gardeners. Rainwater collection provides a simple and low-cost means of procuring and storing water in situ. If no neighboring building or tool shed is on-site to collect water from, however, some other collection surface must be used. One technique developed by the Radix Center was to turn an old satellite dish into a portable rain collection bowl that could be used to provide water to a lot of gardens throughout the city or in otherwise remote locations (Figure 4.7). The following section describes the process of constructing a mobile rain collection system.

Figure 4.7 Simple Rain Barrel

Source: Scott Kellogg

What you will need:

One old satellite dish 6′ to 8′ in diameter (solid surface, not mesh)

- Four 1″-diameter galvanized water pipes cut to 6' lengths
- Four 1″ threaded galvanized "floor flanges" for metal pipe
- 16 1/4″ bolts, 2″ long
- Screw gun with a ¼″ drill bit
- ¼″ socket wrench

What to do:

Step 1: remove any old mounting hardware from the satellite dish. Old dishes can often be obtained for free or low cost – as a commercially obsolete technology they have little more than scrap value.

Step 2: measuring from the outside of the dish, choose four equally spaced locations to attach the floor flanges. Place them roughly halfway between the center and the outside of the dish – this will ensure stability of the dishes. Using a pencil or marker, mark the locations of the flange screw holes on the surface of the dish.

Step 3: using the screw gun, drill holes in each of the marked spaces and bolt the flanges to the underside of the dish.

Step 4: turn the dish on its side and carefully screw in each pipe to the flanges – these will act as legs for the dish to stand on. When finished, stand the dish upright.

Step 5: drill several holes in the very bottom of the dish – these will allow water to drain out.

An open-top fifty-five-gallon barrel is placed directly under the dish. As it collects rain, it will fill the barrel beneath, greatly increasing the catchment area that the barrel alone would have. The entire set up should be placed away from any trees, as fallen leaves can quickly clog the drainage holes (Figure 4.8). The system should also be taken down before heavy snowfall, which could collapse it. The dish can easily be stored for the winter by flipping it upside down where it stands.

Oyster remediation

Oysters are bivalves, a type of organism that includes clams, mussels, and scallops. They inhabit the salty and brackish waters that are found in estuaries and coastal areas upon which cities have been built. In addition to playing critical roles in aquatic ecosystems, oysters have been an important food source for humans for millennia. Archeological digs along waterways in New York City have revealed numerous middens, or mountainous piles of shells, that serve as a testament to the importance of oysters as a food source to coastal communities. With calcium being a major mineral component of oyster shells, particularly large-shell middens have altered the chemistry of the soils in their vicinity, creating an environment that

Figure 4.8 A Completed System

favors calcium-tolerant plants and an example of pre-industrial human's ability to alter the biochemical makeup of their environment (McAvoy, 2012). Historically, 25 percent of New York Harbor used to be covered by oyster beds, and twelve-inch wide oysters were once abundant in the Hudson River Estuary and New York Harbor. The abundant shellfish populations served as an important food source to New York City, with its once ubiquitous oyster vending street carts contributing to the city's moniker "The Big Oyster" (Kurlansky, 2007).

Oysters have been described as ecological engineers (Grabowski, 2007). They have a significant effect on the environments in which they live, performing numerous ecological functions including sculpting and stabilizing underwater terrain and purifying waters. Oyster beds, or reefs, are conglomerations of living oysters growing on top of the shells of dead oysters. Some of these reefs grow quite large, both laterally, and to a lesser degree vertically. These beds act like anchors in the mud, halting the erosion of aquatic sediments, much like how the roots of a tree hold soil together and prevent it from washing away. In this regard, the beds stabilize the banks of intertidal zones in mudflats, places that are submerged at high tide and exposed when the tide is low. This oyster shell city is a refuge for biodiversity, creating habitat for an array of species including snails, barnacles, eelgrass, anemones, crabs, and fish. The enormous amount of surface area made by the shells also creates an ideal ecosystem for microbial colonization.

Much like corals do, oysters create their calcium carbonate shells and reefs by precipitating minerals out of the water column. Because calcareous corals are not found outside of tropical zones, oysters fill an important ecological niche in subtropical and temperate waters. Like coral reefs, oyster reefs can act like an offshore buffer to wave action, slowing water down and creating a more protected environment along the shore. Restoration of oyster reefs can be an important strategy for mitigating the effects of coastal storms and hurricanes, projected to intensify along with rising ocean levels because of global climate change.

Oysters are filter feeders, which means they eat by pulling water through their shells and removing algae, organic material, and other nutrients. The suspended particles and nutrients that they feed on are generally regarded as water pollutants, many of which contribute to eutrophication and water turbidity, or cloudiness. Acting like living filters, oysters pull water through their shells, extract nutrients and expel fecal pellets. Removed from the water column, excess nutrients are further broken down by anaerobic bacteria residing in tidal flats. Oysters can filter water at an impressive rate – a single oyster can cycle approximately fifty gallons of water per day through its body (Pelton, 2010).

Unfortunately, oyster populations have suffered dramatically in recent decades. The combined effects of land-based erosion, water pollution, overharvesting, and introduced disease have caused native oyster populations to plummet. Globally, over 85 percent of oyster reefs have been destroyed (Beck, 2011).

In response to the bleak prognosis of oyster populations, a handful of organizations comprising citizen-scientists have taken it upon themselves to restore oyster reefs, with some focusing on urban environments. The Oyster Restoration Research Partnership (ORRP) is one such example, which is attempting to reestablish oyster reefs in several locations in New York City waters (Grizzle, 2013). There are also numerous locations along the Chesapeake Bay, in the watershed of Baltimore, where people have created oyster nurseries attached to docks and floating platforms. For many, the activity of oyster reef restoration reinforces a sense of socio-ecological memory and connection, fostering a greater sense of place and relationship to their environments (Krasny, 2014).

It is important to point out that even if oyster restoration efforts in urban waters are successful, it still would not be safe to eat them. Oysters grown in polluted waters are known to harbor pathogenic organisms that can be passed on to humans who eat them. The primary sources of pathogens are sewage overflows. So, while oysters can do enormous good for regenerating the health and cleanliness of waters, their restoration must go hand and hand with efforts to stop CSOs and to improve the condition of urban stormwater runoff. These strategies combined are critical for building a healthy urban aquatic ecosystem in coastal cities.

Oysters make an interesting topic for framing conversations around urban ecosystem justice, if only for cities adjacent to brackish and saline waters. Although they have been largely extirpated from urban ecosystems in the past century, oysters represent a species that can be thought of as being "native" to the urban ecosystem and have a strong potential to connect urban residents to the health and well-being of urban watersheds. Through their long historical use as an important food source for coastal populations, as a key economic resource, and for the critical role they play in ecosystem engineering and water purification, there already exists strong precedents for furthering mutually reciprocal and symbiotic relationships between humans and oysters in the urban ecosystem. As oysters have historically, and could potentially again in the future, serve as an important food source for lower-income urban residents, they have and could assist in addressing issues of food security and access. Finally, as water purifiers, oysters can help to reduce exposure to waterborne toxic pollutants that many waterfront environmental justice communities suffer from, also helping to reduce concentrations of toxicity in fish, an important source for many low-income urban residents.

Aquaponics as a transitional strategy for restoring urban fisheries

While much of the fish and aquatic life in urban waterways has accumulated dangerous quantities of toxins rendering them unfit to eat, it may be possible through a method called "aquaponics" to provide sustainably produced, locally grown, and nontoxic fish protein to urban residents. There is a fine line, however, between aquaponics being a complimentary component of integrated sustainable food production systems and being an energy-intensive techno-fix form of farming method that furthers the nature–society divide. Aquaponics should only be thought of as an interim strategy to be carried out in tandem with the long-term work of restoring the health of urban fisheries, less it be appropriated by enviro-technocrats and result in increased ecological alienation.

Aquaponics is a method of "soil-less" growing that cultivates both fish and plants together in an integrated recirculating system. Within it, fish wastes are delivered through water to the roots of edible plants, whom they nourish. The plants and their attached microbial community in turn purify the water, which is then returned to the fish, completing the cycle. In this sense, the plants act as a biological filter of fish waste, eliminating the need for synthetic filtering media or nutrient-laden discharge associated with conventional aquaculture. Because the fish feed and water inputs going into an aquaponics system are controlled, it is conceivably possible to produce aquatic foods with concentrations of toxicity dramatically lower than those that would be found in fish taken from an urban river.

A fantastic teaching tool, an aquaponics system demonstrates the concept of cyclical ecosystems, modeling how aquatic environments function. Through it, links can be created to discussion topics ranging from the pollution of urban waterways to humanity's impact on oceanic health. Aquaponics is also highly scalable. The same basic principles can be applied to systems ranging in size from a five-gallon window unit to a giant system contained inside a warehouse. For these reasons, aquaponics can have applicability as a food production technique to landless urban residents.

While there is currently a great deal of enthusiasm about the potential of aquaponics among urban farmers and sustainability advocates, it must be tempered by a realistic assessment of the technology's shortcomings and limitations. To begin, if not done correctly, fish farming can have a just as negative or worse impact on the health of oceanic fisheries than extractive fishing practices. To illustrate, farm-raised fish are largely fed a diet of processed fish protein. The source of this protein is frequently the pelletized remains of by-catch, the unwanted marine creatures caught through commercial fisheries. It is estimated that two to five pounds of protein from wild-caught fish are required for every pound of weight gained by a farm-raised salmon (Barcott, 2001). On account of this, eating commercially farmed fish can contribute greatly to the decline of already imperiled oceanic health. While it's possible to construct food chains of sustainable proteins derived from terrestrial sources for fish (soldier fly larvae, duckweed, mealworms, etc.), it is considerably more difficult when compared to raising chickens or other microlivestock, which may be reared entirely from human-generated food wastes.

Furthermore, aquaponics uses considerably more energy than traditional soil-based gardening as it requires nearly constant mechanical aeration and filtration with electric pumps. While these could conceivably be powered from renewable energy sources, there is still considerable embodied energy invested in the equipment's manufacture. Also required are impermeable materials such as plastic tanks and rubber pond liners, whose construction has significant environmental impacts. Lastly, aquaponics demands a high degree of technological knowledge and expertise from users that greatly exceeds what is comparatively required for soil-based vegetable farming (Figure 4.9). These factors combined will greatly limit its ability to address food provisioning on a broad scale and decentralized level. Accordingly, aquaponics should only be complimentary to rather than a replacement for traditional soil-based gardening. This awareness is important to emphasize as aquaponics, hydroponics, vertical farming, and other controlled environment and resource-intensive food growing techniques are being advocated for as technological solutions for food production in the face of the unpredictability of climate change by such groups as the Ecomodernists (Asafu-Adjaye, 2015); (Despommier, 2013). The techno-environmental agenda espoused by these schools of thought are deceptively "green" manifestations of neoliberal economics and policy. Care must be taken to ensure that aquaponics remains a decentralized technology that enhances, rather than separates, urban residents' relationships to their waters.

Figure 4.9 Aquaponics at Radix
Source: Scott Kellogg

Urban stream restoration

Another urban waterway justice issue closely related to CSO prevention is that of urban stream restoration. Among the biggest threats to urban stream health, a distinct body of environmental justice themed literature that analyzes urban streams specifically has been written and is deserving of review.

Every city has been built upon land that at one time had streams, creeks, and rivers running across it, and for the most part, these natural patterns of topography and drainage still exist and shape urban hydrological processes. Water will still flow to lower elevations, draining into streams that are either exposed or concealed below ground. To urban planners in the nineteenth and twentieth centuries, streams and rivers flowing through cities were undesirable. Used frequently as open sewers, waterways were odiferous, posed public health issues, and impeded development. The solution employed by engineers was to either redirect streams outside of cities or, more commonly, bury them underground. Often, this process of stream burial was gradual: bridges were first built across rivers and were followed by streets, warehouses, and other structures. Eventually, urban streams would be covered entirely, laid in concrete culverts and hidden from public view. Today, many urban residents are completely unaware of the existence of the creeks and streams that still flow beneath their feet. Those that remain above ground are commonly "straightened": channeled into fenced-off concrete sluices that rapidly expel water from cities and create a hazardous and inaccessible stream shoreline (Buchholz, 2007).

Burying streams not only results in the erasure of a city's ecological history, it also worsens the quality of the waters flowing through them. Ecologically intact streams are lined by riparian zones: wetlands that act like buffers along a stream's banks. Riparian buffers trap and degrade surface pollutants before they make their way into water and exacerbate environmental harm and provide numerous ecosystem services including providing habitat to an array of species. When streams are relegated beneath the earth, their associated riparian buffers are destroyed, and the benefits of natural pollutant filtration are lost. Natural streams are also characterized by numerous rocks, submerged logs, falls, pools, bends, and a generally meandering course, all of which help to slow the speed of water and create habitats for a variety of organisms along the streams' "edge zones". When streams are "straightened" and directed through smooth concrete pipes, they increase in velocity and in their erosive capacity. When fast moving water is ejected from culverts and flows across land, it washes away topsoil, uproots trees, damages roads, and can trigger landslides (Neale, 2016).

To rectify the damage caused through stream burial, a practice called stream "daylighting" has begun. Daylighting involves uncovering old urban streams and bringing them back to the earth's surface where they are once again exposed to the "daylight". The environmental benefits of daylighting are many. As the name implies, the quality of water can be greatly improved just by exposing it to air and sunlight. The ultraviolet radiation in sunlight, a natural disinfectant, can kill many of the harmful pathogens that enter streams through urban runoff. Furthermore, by recreating a natural streambed (a common component of daylighting projects), water is slowed and its potential to cause erosion and flooding is reduced. A permeable stream bottom gives stormwater the opportunity to infiltrate soils and reduces the tendency to flood. Restoration of the soil, microorganisms, and plant community along a streambed helps to filter urban runoff from entering the stream. Additionally, a restored stream can create a vibrant wildlife corridor through the heart of a city. There are considerable economic benefits to daylighting as well. Buried stream culverts often become clogged and broken. Bringing culverts above ground and creating a more natural streambed can cost just as much as a conventional repair. Lastly, exposing streams reconnects urban dwellers with their watershed, fostering "ecological memory" (Barthel, 2010) acting as a conduit to the environment beyond the city limits. Daylighted creeks can be transformed into urban parks, creating a space for congregation and connection around an ecological feature.

Daylighting and stream revitalization offer tremendous potential to revitalize urban ecosystems and engage surrounding communities. Unfortunately, high costs create a significant barrier to stream restoration. The recent daylighting of the Sawmill Creek in Yonkers, NY had a budget requiring a thirty-four-million-dollar state grant. There are a handful of studies that examine the explicit environmental justice implications of urban stream restoration. One finding has been that most stream restoration occurs in rural areas, leaving low-income urban populations spatially disenfranchised (Moran, 2010). The effects of the unequal geographic distribution of stream restoration projects go beyond low-income urban communities merely being excluded from the benefits of stream restoration. In addition, communities living adjacent to urban streams must additionally continue to bear the brunt of exposure to the toxins present in un-restored streams. Furthermore, the process and industry of stream restoration itself has become dominated by neoliberal interests and "the tyranny of relevance", further reducing the possibility of low-income communities benefiting from restoration efforts (Lave, 2016). Even when restorations do take place in lower-income urban environments, great care must be taken to ensure the process is participatory and considers the concerns and needs of residents (Moran, 2010), else the communities living in the vicinity of the restored stream continue to feel alienated by it.

One way that citizens can actively work to build popular support for daylighting projects is first to raise people's awareness of the existence of underground streams. Historical maps will often show waterways and streams, offering clues to their current location (in many cases still flowing through the same areas). Mapping projects provide an engaging way to overlay the historic paths of streams on top of their current courses. Spray painting stencils on streets and sidewalks to mark the path of underground streams create a way to bring attention to the existence of streams and to help in connecting urban residents to ecological place. Such a project was carried out at the Radix Center by local high school youth who mapped both the historical and present-day locations of the Beaverkill Creek, a tributary of the Hudson that flows beneath Lincoln Park and directly alongside the Radix Center. Youth were further engaged by locating the confluence of the Beaverkill and the Hudson and doing water quality analysis of this spot.

Conclusion

The problems facing urban waterways are genuinely wicked, some of which can only be solved through multimillion dollar infrastructural upgrades performed by municipalities. How then, is it possible for the average urban resident to have any impact on this problem? The good news is that there are several simple, affordable, and decentralized techniques that can be carried out, which will result in mutual and reciprocal benefit to both people and the health of their waterways. While the impact that any of these might have by themselves is small, collectively and synergistically they can produce significant results. It is critical to point out that these decentralized approaches must be done in coordination with broader political actions aimed at "turning off the tap" of stormwater and sewage pollution – residents cannot bear the burden of "mopping up the mess". More importantly, citizens taking any kind of initiative toward improving their relationships with local waterways have profound symbolic and educational value. Just thinking of impaired waters as anything other than hopelessly polluted and deserving of care creates a powerful counter narrative to the idea of urban waters as being dead, toxic, and beyond salvation – an essential first step toward building urban ecosystem justice and the restoration of the aquatic commons.

Works cited

Asafu-Adjaye, J. B. (2015). *An ecomodernist manifesto.* Oakland, CA: Breakthrough Institute.

Barcott, B. a. (2001). Aquaculture's troubled harvest. *Mother Jones*, *26*(6), 38–45.

Barthel, S. C. (2010). Social – ecological memory in urban gardens: Retaining the capacity for management of ecosystem services. *Global Environmental Change, 20*(2), 255–265.

Beatley, T. (2014). *Blue urbanism: Urban design for a blue planet*. Washington, DC: Island Press/Center for Resource Economics.

Beck, M. W. (2011). Oyster reefs at risk and recommendations for conservation, restoration, and management. *Bioscience, 61*(2), 107–116.

Borchardt, D. a. (1997). Urban stormwater discharges: Ecological effects on receiving waters and consequences for technical measures. *Water Science and Technology*, 36(8), 173–178.

Buchholz, T. a. (2007). *Urban stream daylighting: Case study evaluations*. Blacksburg, VA: Virginia Water Resources Research Center.

Bullard, R. D. (1999). Atlanta megasprawl. *Forum for Applied Research and Public Policy, 14*(3).

Cao, H. e. (2010). Heavy metals in rice and garden vegetables and their potential health risks to inhabitants in the vicinity of an industrial zone in Jiangsu, China. *Journal of Environmental Sciences (China), 22*, 1792–1799.

Cao, W., & Zhang, Y. (2014). Removal of nitrogen (N) from hypereutrophic waters by ecological floating beds (EFBs) with various substrates. *Ecological Engineering, 62*, 148–152.

Checker, M. (2018). Green apple, brown spots: Environmental gentrification in New York City. In *Urban life: Readings in the anthropology of the city* (p. 413). Long Grove, IL: Waveland Press.

Chen, G. a. (2010). Penetration of cover crop roots through compacted soils. *Plant and Soil, 331*(1–2), 31–43.

Curran, W. a. (2018). A just enough green? Industrial gentrification and competing socionatures in Greenpoint, Brooklyn. In *Just green enough* (Vol. 32, pp. 32–46). London: Routledge in association with GSE Research.

Danks, S. G. (2010). *Asphalt to ecosystems: Design ideas for schoolyard transformation*. New York: New Village Press.

Despommier, D. (2013). Farming up the city: The rise of urban vertical farms. *Trends in Biotechnology, 31*(7), 388–389.

Flannery, T. (2002). *The future eaters: An ecological history of the Australasian lands and people*. New York: Grove Press.

Frazer, L. (2005). Paving paradise: The peril of impervious surfaces. *Environmental Health Perspectives, 113*(7), A456.

Grabowski, J. H. (2007). Restoring oyster reefs to recover ecosystem services. In *Ecosystem engineers: Plants to protists* (pp. 281–298). London: Elsevier.

Grizzle, R. e. (2013). *Oyster Restoration Research Project (ORPP) final technical report: ORRP Phase 1: Experimental oyster reef development and performance results*. New York: US Army Corps of Engineers – New York District, The Port Authority of New York & New Jersey.

Guoyan, C. H. (2005). Treatment in water bodies pollution by ecological floating bed technology. *China Water Kesources, 5*, 22.

Ingham, E. R. (2000). Brewing compost tea. *Kitchen Gardener, 29*, 16–19.

Krasny, M. E. (2012). Civic ecology: A pathway for earth stewardship in cities. *Frontiers in Ecology and the Environment, 10*(5), 267–273.

Krasny, M. E. (2014). New York City's oyster gardeners: Memories and meanings as motivations for volunteer environmental stewardship. *Landscape and Urban Planning, 132*, 16–25.

Kurlansky, M. (2007). *The big oyster: History on the half shell*. New York: Random House.

Lave, R. (2016). Stream restoration and the surprisingly social dynamics of science. *Wiley Interdisciplinary Reviews: Water, 3*(1), 75–81.

LESEC. (2015). *Lower east side ecology center*. Retrieved December 11, 2015, from www.lesecologycenter.org

Logsdon, G. (1978). *Getting food from water: A guide to backyard aquaculture*. Emmaus, PA: Rodale Press, Inc.

Lundy, L. a. (2011). Integrating sciences to sustain urban ecosystem services. *Progress in Physical Geography*, *35*(5), 653–669.

McAvoy, W. A. (2012). Plant community classification and the flora of Native American shell-middens on the Delmarva Peninsula. *Maryland Naturalist*, *53*(1), 1–34.

McLarney, W. O. (1984). *The freshwater aquaculture book: A handbook for small scale fish culture in North America*. Washington, DC: Hartley and Marks.

Moran, S. (2010). Cities, creeks, and erasure: Stream restoration and environmental justice. *Environmental Justice*, *3*(2), 61–69.

Neale, M. W. (2016). Re-engineering buried urban streams: Daylighting results in rapid changes in stream invertebrate communities. *Ecological Engineering*, *87*, 175–184.

Noonan, D. S. (2005). Neighbours, barriers and urban environments: Are things different on the other side of the tracks? *Urban Studies*, *42*(10), 1817–1835.

Pelton, T. a. (2010). *On the Brink-Chesapeake's native oysters: What will it take to bring them back?* Chesapeake Bay Program. Retrieved from https://mdstatedocs.slrc.info/digital/collection/mdgov/id/0/

Porter, M. J. (2015). The Hudson River watershed, New York State, USA. In *Catchment and river basin management* (pp. 95–117). London: Routledge.

Sedlak, D. (2016). Fool me once. *Environmental Science and Technology*, *50*, 7937–7938.

Sloan, R. J. (1988). PCB patterns in Hudson River fish: II: Migrant and marine species. In *Fisheries research in the Hudson River* (p. 325). Albany, NY: State University of New York Press.

Tsihrintzis, V. A. (1997). Modeling and management of urban stormwater runoff quality: A review. *Water Resources Management*, *11*(2), 136–164.

Wattigney, W. A.-W. (2017). The biomonitoring of great lakes populations program. *Journal of Environmental Health*, *79*(8), 42.

Willems, P. e. (2012). Climate change impact assessment on urban rainfall extremes and urban drainage: Methods and shortcomings. *Atmospheric Research*, *103*, 106–118.

Zhao, F. e. (2012). Purifying Eutrophic River waters with integrated floating island systems. *Ecological Engineering*, *40*, 53–60.

5 Urban atmospheric justice

Breathing air is the most essential act of life and living, with the atmosphere serving as an interconnecting medium of a respiring planet. Vapor, microbes, dust, gases, pollen, smoke, all pass through our lungs to our blood and again circulate world-round. Breathing trees transfer moisture from the ocean far inland in the form of "flying rivers" of clouds, creating forests where otherwise there would only be desert (Sheil, 2018). The "thin blue line" of the earth's atmosphere serves as a fragile buffer between us and the hostility and coldness of space. While food and water can be forgone for days, it is impossible to survive more than a few minutes without air. Despite this, invisible and minimally tangible air is often taken for granted – we only appreciate its criticality when we are deprived of it (Nieuwenhuis, 2018). Consequently, atmospheres are too often used as "landfills in the sky", polluted with emissions from industry, transportation, and incineration. This disregard for air quality is responsible for an estimated one in nine deaths each year, making it one of the greatest environmental threats to human health (World Health Organization, 2016). Even climate change, arguably, is symptomatic of our blindness to and unreceptivity to air – the very medium in which we exist – as we wantonly unbalance our atmosphere with heat-trapping fumes (Abram, 2018). Air pollution and climate change negatively synergize in urban areas where vehicular emissions and the urban heat island effect compound to disproportionately impact vulnerable populations (Kumar, 2015). "I can't breathe", the last words of Eric Garner before being suffocated by police, has become a rallying slogan of the Black Lives Matters movement, representing a politicization of the very right to breathe (Górska, 2016). In light of these injustices, a new relationship between air and human culture is desperately needed which centers both the atmosphere's critical ecological function and the essential human right of healthy air.

To cultivate greater respect and reciprocity between humans and the atmosphere, it can help to conceptualize air as a living system. Although it is typically thought of as sterile, air is in fact teeming with life. The discipline of aeroecology is dedicated to the study of the flying, or "volant", organisms that use the aerosphere as habitat and for migration – birds, bats, insects, and spiders (Chilson, 2017). Yet another field of study, aerobiology, has shown that there are an estimated 10^{19} microbes living in the earth's clouds – a rich ecosystem of bacteria, fungi, and algae protected from the extreme conditions of the upper atmosphere by the floating

DOI: 10.4324/9780367858711-6

blankets of moisture in which they reside (Amato, 2012). These aerial organisms play significant roles in planetary processes, including creating and influencing weather patterns. *Pseudomonas syringae*, a cloud-borne bacterium, manufactures proteins that make it an ideal nucleus for ice formation. Falling back to the earth as snow or other frozen precipitation, it lands on plant leaves, infecting them through the freeze-damage wound. After feeding on the plant, the leaves will dry, become dust, and winds will carry the bacterial spores back up into the clouds, completing the cycle (Ogden, 2014). Similarly, marine algae produce a gas called dimethyl sulfide that aids in cloud formation. These biogenic clouds help cool earth temperatures and serve to disperse the wind-lofted algae across the ocean (O'Dowd, 2002). Combined, these two aerobiological processes provide evidence for the biosphere's Gaian capacity for self-regulation (Hamilton, 1998). While the aerosphere all and all is less abundant in life than water and soil, it is still very much a part of the living earth, and its well-being is highly interconnected with that of humanity.

The political–ecological dimensions of air need to likewise be appreciated and how rights to clean air are disproportionately granted to privileged populations. Sociologist Ulrich Beck stated that "Poverty is hierarchic, smog is democratic" (Beck, 1995), referencing how wind-blown pollutants crossing neighborhood lines impact the rich and the poor alike. Air is a fluid, mobile, transboundary element. Defying containment, air freely transgresses the demarcations of brownfield exclusion zones and walled-off superfund rivers that encapsulate contaminated soil and water. Redlining policies intended to confine undesirable social and ecological factors cannot impede the drift of ozone and ultrafine particulates into the airsheds of affluent downwind suburbs or block blankets of smog from enveloping both slums and penthouses (Manigrasso, 2017). On a global scale, air pollutants permeate national borders freely, with Indonesian forest fires and Gobi Desert dust storms impacting communities on the opposite side of the world (Negi, 2020). The distribution of toxic air is not entirely egalitarian, however. Increasingly, we are aware of the extent to which low-income communities, through their proximity to pollution sources, bear the brunt of exposure to air pollution (Morello-Frosch, 2001). While lighter air pollutants can be carried great distances on the wind, the heavy particulate fallout of highways and factories concentrates close to the ground in low-lying areas, impacting pejoratively labeled "bottom dwellers". Urbanites with the means live in high-rise apartments above the filth, creating a real estate market that is literally vertically stratified (Graham, 2015). Coupled with air-conditioning, a luxury of the wealthy classes through much of the world, the rich live in filtered-climate controlled indoor environments, insulated and isolated from the eco-social unpleasantries of the world outside. As the world warms, air-conditioning – itself a primary driver of climate change (Hawken, 2017) – is used more and more, creating a positive feedback loop of urban warming and inequality (Takane, 2019). When industrial emissions, traffic pollutants, biomass burning, dust storms, trash fires, fireworks, and teargas converge with the right climatic and geographic conditions, toxic air domes can form over cities, leading to emergent events such as the widely reported "Airpocalypse" (Ghertner, 2020) – nothing short of atmospheric terrorism, or an "atmoterrorism" (Sloterdijk, 2009) of late capitalism upon the aerial commons.

Air governance

From an angle of governance and management, the sources of air pollution are diffuse and harder to pinpoint – air is all around us, inescapable, permeating our tissues and blood. Contaminated soils and rivers can be encapsulated, covered, and avoided by abstaining from activities such as gardening, swimming, and fishing. Ubiquitous air, however, is far more difficult to hide from. Sources of water pollution, such as sewage overflows and stormwater runoffs, are relatively localized and concentrated – the result of multiple nonpoint sources converging to a single stream in concrete culverts discharging into waterways. This confluence makes it easier, at least conceptually, to contain, to dam the flow. Air pollution dispersal patterns, in contrast, are nearly the inverse of those of sewage pollution. While air also comes from multiple sources, rather than being funneled to a single point, it is instead dispersed broadly like a spreading river delta or a branching tree. Combined with its intangibility and frequent invisibility, this makes air harder to pinpoint, isolate, and confine. While air quality certainly has improved in the United States on account of the Clean Air Act and other policy measures (Chay, 2003), air pollution continues still, remaining diffuse, elusive, and systemic. To take on the combined sources of air pollution is to challenge core economic processes and cultural practices in our world. Fossil fuel extraction, manufacturing, auto dependence, urbanization, the growth economy itself – these are all comparatively far more difficult to redesign than ill-adapted sewage infrastructure. Local solutions are of some, but limited, effectiveness for improving air quality. While attempts at international coordination of air quality governance have been made in the form of the Convention on Long-Range Transboundary Air Pollution, or CLRTAP (Maas & Grennfelt, 2016), greater cooperation is required to confront this global threat.

In order to protect themselves from air toxicity, some have taken to wearing gas masks or respirators outdoors when exercising or even biking to work along roadsides. These, it is argued, are tools of survival. "Means of air conditioning (e.g. gas masks) have instead turned into the primary instances of resilience. . . . Any form of effective resistance against atmo-terrorism should, therefore, start by securing the air from which life springs" (Nieuwenhuis, 2013). While a gas mask is lifesaving in a tear-gas choked demonstration or a chemical accident, and as such is symbolic of resistance, it is also a symbol of war, industrialized terror, and ecosystemic collapse. Along with other breath-enhancing techno-fixes like "oxygen bars", gas masks are representative of the privatization, commodification, and devastation of the atmospheric commons. We are in dire times when it becomes necessary to wear a gas mask to breathe outside of the home.[1] Individual protective gear is insufficient for challenging the systemic and political drivers of poor air quality.

Resilient airsheds require that environmental justice activists hold governmental agencies accountable for aggressively regulating air pollution. In the interim, and as a parallel strategy, can anything be done on a local, community-based level to protect vulnerable populations from air toxins?

Air remediation

Compared to soil and water, urban air and climate is also more difficult to address through a framework of urban ecosystem justice, at least in the sense of there being options for actionable citizen-based bioremediation and mutualistic human–ecosystem reciprocity. Soil contaminants can be degraded and leave you with quality garden soils in exchange. Rainwater can be captured, stored, and purified, giving you water for drinking, irrigation, or aquaculture. In these examples, simple and affordable techniques and practices result in the empowerment of individuals and communities. While these actions may not result in broadscale systemic transformation, they nonetheless provide hope that when scaled up, individual actions can have a collective synergistic impact. Fewer options exist for air.

Very broadly, air remediation can be thought of in two distinct categories: indoor and outdoor air. While outdoor air is difficult to contain, isolate, and purify, the potential of affordable indoor air remediation, closed environments where aerial exchanges with the outdoors are limited, is promising.

In addition to increasing oxygen levels and creating aesthetic beauty, common houseplants have demonstrated the ability to remove airborne toxins in indoor spaces. B.C. Wolverton's book "How to Grow Fresh Air" discusses the use of houseplants to purify enclosed environments – a method researched by NASA for use in the sealed environment of a spacecraft to detoxify the poisonous "bioeffluents" of the human body as well as the volatile organic compounds (VOCs) given off by the space capsule materials (Wolverton B. C., 1997). Through transpiration, plants draw air through their roots where contaminants are trapped and deactivated by associated microbes. In addition to toxic bodily emissions, a wide range of indoor air pollutants including benzene, toluene, octane, trichloroethylene, and formaldehyde (Yang D. S., 2009); (Kim, 2010), many of which are off-gassed from building materials and products in the home. In an unpublished study (Wolverton B., 2010), Wolverton demonstrated the ability of fan-assisted houseplants to significantly reduce concentrations of airborne formaldehyde in the infamous "Katrina trailers" that were widely distributed to survivors of Hurricane Katrina in Louisiana by FEMA despite containing dangerous levels of formaldehyde gas (Murphy, 2013). Showing great potential to be used in similar environmental justice applications, indoor houseplant air purification technology is simple, affordable, low-energy, and can be scaled-up for whole building "green wall" systems (Irga, 2018). It's important to mention that plant-based air purification technology is still being studied to maximize its effectiveness – passive diffusion of air through plants' rhizospheres may not be sufficient to adequately reduce indoor air pollutants, and some component of active ventilation may be required (Waring, 2016). Despite critiques, there is a multitude of scientific studies to support the efficacy of using houseplants for indoor air purification (Aydogan & Cerone, 2021), and for front-line environmental justice communities, few other affordable options may exist.

According to Wolverton, the top ten plants for indoor air purification include:[2]

(1) Areca Palm (*Chrysalidocarpus lutescens*)
(2) Lady Palm (*Rhapis excelsa*)
(3) Bamboo Plant (*Chamaedorea erumpens*)
(4) Rubber Plant (*Ficus elastica*)
(5) Dracaena (*Dracaena decremensis*)
(6) English Ivy (*Hedera helix*)
(7) Dwarf Date Palm (*Phoenix roebelenii*)
(8) Ficus (*Ficus macleilandii*)
(9) Boston Fern (*Nephrolepsis exaltata*)
(10) Peace Lily (*Spathiphyllum wallisii*)

(Wolverton B. C., 1997)

What happens, however, when we go outside of air-conditioned or phyto-filtered indoor environments? Outdoor air is a different question. Rapidly flowing, rising, falling, air masses cycle in and out as quickly as the wind changes direction – or sometimes they don't – heat inversions and stagnant breezes trap bubbles of caustic air in urban valleys for days as smog, particulates, and ozone levels escalate. What, if anything, can be done on a citizen science, community-based, DIY level to purify polluted outdoor air, reduce the damage it causes, and monitor and raise awareness about air quality? Urban ecosystem justice advocates a combined strategy that includes citizen-centered air monitoring, education, and the creation of vegetated depositional barriers around sensitive areas and vulnerable populations.

Air pollution basics

There are several types of urban air pollutants: these include noxious gaseous pollutants such as carbon monoxide (CO) and nitrous oxides (NOx) (Barwise, 2020). There is also particulate matter (PM), which is classified by size. PM_{10} are particles ten micrometers or smaller, while $PM_{2.5}$ are particles two and a half micrometers or less (Figure 5.1). The latter are of particular concern to human health as they may penetrate deep into the lungs and enter the bloodstream (EPA, 2020a).

Additionally, there are ultrafine particles (smaller than 0.1 micrometers), a highly mobile pollutant whose effects on health are concerning yet relatively understudied (Kumar, 2014).

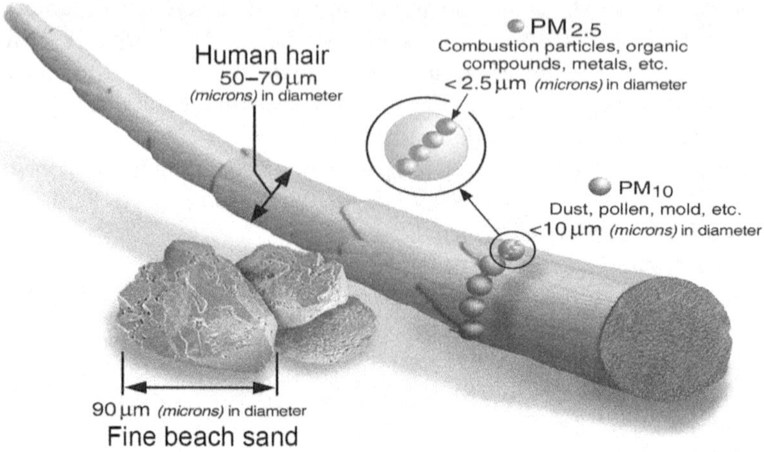

Figure 5.1 Size Comparison of Air Pollutants (EPA, 2020b)

Air monitoring

Historically, and today, government-run air quality monitoring stations are few and far between, with perhaps one or two operating within smaller metropolitan regions. This has been justified by governmental agencies according to their belief that air quality patterns are relatively uniform across regions and that data collected in a suburban region would accurately reflect conditions in an urban center. Recent studies have shown, however, that air pollution patterns can be highly localized, with particulate emissions and even ozone varying considerably between neighborhoods accordingly to their proximity to pollution sources (Houston, 2015). Findings from these studies would suggest a need for a greater number of air quality monitoring stations more widely dispersed through a city to record the variations in local conditions.

In the past, cost has been a significant barrier to building more monitoring stations. Recently however, higher quality air quality monitoring technology has become considerably more affordable. There are now a number of low-cost commercially available hand-held air quality monitoring devices that, despite broader concerns over their reliability and the value of citizen-derived data (Ottinger, 2010), allow large numbers of citizens to conduct multiple informal air quality readings at widespread locations (Mahajan & Kumar, 2020). These low-cost sensors often employ open access software and communications technologies that can network their readings and produce fine-grained information. For this reason, the citizen-science potential of air quality monitoring is enormous, with its decentralized accessibility and participatory appeal grounding and familiarizing the otherwise highly diffuse and intangible nature of air. Studies related to citizen-science air monitoring are many (Kaufman, 2017); (Bo, 2020); (Golumbic, 2019); (Dutta, 2009); (Chen, 2017); (English, 2017). The best among them emphasize the importance of the citizen science process being inclusive (involving historically underrepresented and marginalized communities), collaborative (promoting interactions between multiple stakeholders including researchers, policymakers, and communities so as to further community capacity building) and reciprocal (actively debating results and outputs) (Mahajan, 2020). The continuing exploration of citizen-driven air quality monitoring is an integral component of urban atmospheric justice.

A second option would be to make existing air quality data more accessible and visible to a greater number of people. While Air Quality Index (AQI) information is readily available online, the "digital divide" (Ragnedda, 2017) – the fact of low-income people commonly lacking internet access – makes this information inaccessible to some of the most vulnerable populations. EPA-sponsored programs such as the air quality flag program, where citizen groups would fly colored banners that correspond with the color system used in the AQI, would help to inform a greater number of citizens about current air quality conditions (Shendell, 2007). In the past, the Radix Center has run educational games for elementary school age children based around learning the names and meanings of the air quality index flag colors. These games have tied into an approach to place-based learning (Sobel, 2004) that teaches ecological literacy while acknowledging the prevalence of pollution-related ailments such as asthma.

Vulnerable areas such as apartment units, gardens, playgrounds, and parks that are near air pollution sources could be protected by constructing vegetated depositional barriers. These essentially are hedgerows of trees or vine-covered fencing that serve to block and trap airborne particulates through a process called "dry deposition" – basically pollutant particles become stuck on the surfaces of leaves. The ability of plants and trees to improve air quality and trap pollutants is well documented (Hagler, 2012); (Steffens, 2012); (Ottosen, 2020), although there are many factors which will determine their effectiveness. There is a general recognition that evergreen shrubs provide superior benefits in temperate climates on account of their foliar density and year-round vegetated surfaces (Yang J. Y., 2015). Other factors that enhance a plant's ability to trap pollutants are small size, complex shape (ridges and creases), and having a hairy or waxy surface (Weerakkody, 2018). Construction of depositional barriers is relatively simple and affordable, making them technology usable by urban citizens to provide themselves some line of defense against airborne pollutants. At the Radix Center, we train plants horizontally around our perimeter fence to act as aerial deposition barriers, protecting our interior gardens from airborne toxins. This is important for blocking not only traffic emissions but also urban street dust – clouds containing toxic metals caused by vehicular brake pad wear. While depositional barriers can reduce the amount of toxic metals that land on vegetable surfaces, it is nonetheless important to wash produce prior to consumption in order to remove residual dusts (Tam, 1987).

Urban climate justice

Air and climate are highly interconnected phenomena – it is difficult to discuss one in the absence of the other. Failure to recognize their interwovenness can result in disastrous policy decisions, such as the widespread incentivization of diesel-powered vehicles in London as a purported remedy to climate change in the 2000s (Freytas-Tamura, 2017). While diesel vehicles may get better mileage than gasoline vehicles and consequently release fewer carbon emissions, they also produce considerably more air pollution. London's adoption of a diesel fleet resulted in the unintended consequence of an air quality crisis for the city and represents a classic instance of carbon fundamentalism: where a narrow focus on technological solutions disregards the intersectionality of health, environment, justice, and the root causes of climate change (Dayaneni, 2009).

On account of their elevated temperatures, urban environments are studied as climatological laboratories, modeling the effects of increased temperatures on humans and non-humans that are expected in the rest of the world as a consequence of climate change (Rizwan, 2008). Emerging whole systems analyses such as "Urban Climate Justice" (Steele, 2015) look at the complex relationships between human vulnerability, environmental sustainability, and biodiversity. These equity-oriented frameworks counterbalance carbon fundamentalism and are highly synergetic with the Urban Ecosystem Justice perspective.

Keeping in mind how closely related air and climate are with one another, and that the primary focus of this chapter is on air quality, there are still some more strongly "climatic" aspects of urban atmospheric ecology that warrant their own exploration. Most notable among these is the "urban heat island", the condition where urban environments are as much as seven degrees warmer than rural areas (EPA, Heat Island Effect, 2020), exacerbating the effects of climate change on vulnerable populations. The urban heat island is caused when the immense thermal mass of a city's darkly colored buildings, roadways, parking lots, rooftops, and sidewalks absorb huge amounts of solar radiation in the day[3] and later release it at night. This is added to by the heat continuously being emitted from vehicles and industry within a city. Typically, this heat will be released in the night and be carried away by winds. Given the right meteorological conditions, however, heat-trapping pollution domes, known as inversions, will form over a city and prevents this heat from being dispersed. While this effect can help moderate urban temperatures in the winter,[4] it can also cause cities to become dangerously hot in the summer months, often with deadly consequences.

Events such as heat waves are greatly compounded by the urban heat island effect, with temperatures reaching dangerously high levels, particularly in top-floor housing as it is being baked under black tar rooftops. Especially at risk are the poor and elderly. Many in low-income neighborhoods are unable to afford air-conditioning and, fearing crime, are unwilling to open their windows during a heat wave. The lack of ventilation and cooling can cause an apartment to roast. Such events were responsible for the deaths of thousands in the heatwaves impacting southern Europe in 2003 and Chicago in 1995 (Klinenberg, 2015).

These events again serve to demonstrate how wealth, and the ability to live in an air-conditioned space, can make the difference between survival and death in the urban ecosystem. These disparities will become more pronounced in the future as planetary temperatures continue to climb, rendering certain regions uninhabitable. A recent study has predicted that extreme summertime temperatures in combination with high humidity may cause the climate in Middle Eastern cities to exceed human survivability by the end of the century (Pal, 2015).

The impacts of the urban heat island are inequitably distributed geographically. Various studies have shown how average temperatures within a city will vary considerably on a block-by-block basis, with the strongest determinants being the prevalence of green roof infrastructure and density of tree cover, both of which positively correlate to higher income (Harlan, 2007); (Sanchez, 2019); (Koo et al., 2019). The good news about this is that it is possible for communities to mitigate the urban heat island on a local level by seizing control of their neighborhood's microclimate. Microclimate is defined as the conditions of temperature, humidity, and air flow of a small area that differ from the dominant surrounding climatic conditions. These climatic differences can be substantial, potentially tipping the balance between life and death for vulnerable people. The design of urban microclimate has been extensively studied (Erell, 2012), with interventions increasing the shading, surface albedo, and vegetative cooling of a particular area by replacing paved lots with gardens (Tsilini, 2015), creating vegetated roofs (MacIvor, 2016), digging cooling ponds (Robitu, 2006), and planting street trees. Because they also have significant benefits for air quality and citizen-driven potential, this chapter will focus on tree-planting strategies.

Trees mitigate the effects of the urban heat island in many ways: their green canopies reflect rather than absorb sunrays, they shade the asphalt and concrete surfaces below them, and they cool through evapotranspiration. Additionally, urban trees provide numerous other benefits to people and ecosystems including air purification, biodiversity enhancement, carbon sequestration, soil stabilization, water filtration, food production, and aesthetic beauty (Manes, 2012); (Fares, 2017); (Threlfall, 2017); (Park, 2019). The ecosystem services of trees, however, are not equitably distributed among a city's residents, with low-income communities benefiting less (Nyelele, 2020). It has also been shown how historically redlined neighborhoods have measurably fewer trees today than more highly appraised areas (Locke et al., 2021); (Field, 2020). How can this imbalance be remedied? While some cities have programs for incentivizing and subsidizing tree planting, these are frequently not equitably delivered as well (Garrison, 2018). What measures can communities take to plant trees on their own without governmental assistance, turning grayscapes to greenscapes?

A newly planted street tree must be mature to survive – at least 1¾" in diameter and eight feet tall – tiny 1–2-foot-tall "whips" are likely to be trampled in a highly trafficked sidewalk. Mature trees, however, are expensive – at least one hundred dollars and likely more per tree. When factoring in the additional expenses of removing sidewalk slabs and adding soil amendments, costs can quickly rise if an entire block is to be planted, let alone an entire neighborhood. One strategy that involves time and planning but requires less money is to create a community tree nursery (Wightman, 1999). A nursery allows a community to grow up small trees in pots over a period of years before eventually planting them in the ground. Small trees can be bought in bulk quantities from wholesale nurseries at 1–2 dollars per tree, dramatically reducing the cost of a neighborhood tree planting project. An even more affordable (yet time-intensive) option is to start trees from wild-harvested seeds. A nursery can be established in a backyard or vacant lot, and apart from regular watering, requires little maintenance.

I strongly advocate planting urban street trees that are: (1) (ideally) native to the region where they are being planted, (2) genetically unique, (3) have fruits, nuts, or berries that will provide food to both humans and non-humans.

Native trees, in general, are better adapted to survive in the climatic conditions in which they have evolved. Furthermore, they will support greater levels of native biodiversity through the coevolutionary relationships they have with other species (birds and insects), many of which are obligate (Helden, 2012). With this stated, it is important to keep in mind that on account of climate change, the earth's temperature is rapidly warming, especially in cities. For this reason, it may be wise to plant tree species whose historical ranges are primarily within warmer zones, as they will be more resilient to the higher temperatures of a volatile urban climate (Lohr, 2014). While native trees are mostly preferable, it is important to remember that many commonly eaten edible fruits including apples, peaches, cherries, plums, pears, and so on are not native to North America. The value that trees have in improving community food sovereignty (along with their myriad other social and environmental benefits) must be considered alongside of their biodiversity significance in a balanced assessment of natives versus non-natives selection (Kull, 2018).

Genetic diversity is an important and seldom recognized aspect of urban agroforestry. Urban forests around the world are currently suffering from genetic homogenization, a result of hundreds of cultivar clones being planted. These are not just species monocultures but are stands of genetically identical "twins" – the result of commercial nurseries favoring clonal reproduction techniques that produce desirable, and predictable, traits in every tree. The forests' lack of genetic diversity makes them incredibly vulnerable to disease as the innate resistance that comes from a heterogenous population is lost in this evolutionary dead end (Kendal, 2014). As lower-income neighborhoods have been shown to have lower rates of species diversity in their trees (Avolio, 2018), it is important to correct this imbalance to preserve the ecological integrity of the forests for vulnerable groups. An additional advantage of buying bulk seedlings from wholesale nurseries is that each one, unlike many cultivars available in larger sizes, is genetically unique, having been grown from seed rather than from a tissue cutting.

There is a growing body of literature on the concept of "urban food forests" which examines the role that tree fruit and nut crops can play in building urban food security (Clark, 2013); (McLain R. e., 2012); (Riolo, 2019); (Castro, 2018); (Brito, 2020); (Tellström, 2014). Edible tree crops growing on sidewalks, parks, and vacant lots can provide city residents with a free, nutritious food source made available through "urban foraging" (Fischer, 2020); (Poe, 2014); (Shackleton,

2017); (McLain R. J., 2014); (Landor-Yamagata, 2018); (Bunge, 2019). Foraging is a practice undertaken by urban residents throughout the world to supplement their diets and has been demonstrated to be used in the United States by a wide range of racial groups (Gaither et al., 2020). While there are outstanding concerns about the safety of growing tree crops in contaminated soils, it is believed to be a safer practice than vegetable gardening and as such may hold potential for repurposing brownfield sites (Gori, 2019). Taking advantage of the copious real estate that exists overhead, food producing trees are an example of vertical food production (Beacham, 2019). Tree crops, agroforestry, and food forests all prefer the cultivation of perennial crops that, once established, provide food year after year with relatively minimal maintenance, particularly when compared to annual vegetable cultivation. By simulating the complexity and architecture of a forest, perennial edible systems align themselves with the forces of ecological succession that move fields to shrublands to woodlands. By not resetting the successional clock to zero each year that comes through plowing fields, carbon and organic matter accumulate in the soils around expanding root systems, trees' vertical branching structures complexify, and intricate biocultural diversity relationships unfold (Jacke, 2005). For these reasons, edibility should be a desirable trait in street tree selection (Figure 5.2).

Appropriate species selection will vary considerably from region to region and must be individually researched. Factors of height must also be compatible with urban power line infrastructure. The following is a list of species which Radix uses in Albany, New York, that are both native and edible.

- Serviceberry (*Amelanchier laevis*)
- Paw Paw (*Asimina triloba*)
- American Persimmon (*Diospyros virginiana*)
- Hazelnut (*Corylus americana*)
- Choke Cherry (*Prunus virginiana*)
- Hackberry (*Celtis occidentalis*)
- Butternut (*Juglans cineria*)
- Black Walnut (*Juglans nigra*)
- Honey Locust (*Gleditsia triacanthos*)

S(no)w justice

While much of the focus in this chapter has been on mitigating the effects of the urban heat island in summer and on hot weather-related stresses, it is important to remember that winter is an equally challenging season in cold climate cities throughout the world. Extreme snowstorms have huge impacts on cities, capable of shutting them down for days and bringing all activity to a halt. Inadequate or no home heating creates a hypothermia risk (especially to the homeless), short days with little sunlight contribute to seasonal affective disorder, and impassible roads and sidewalks create dangerous walking conditions and isolation (Haney C., 2020). The environmental impact of snow removal is considerable as well, with the tons of rock salt applied to roads and sidewalks increasing the salinity

Figure 5.2 Community Nursery

Source: Scott Kellogg

of local waterways as it all washes away in the spring (Novotny, 2008). While climate change may be shortening the length of winter and the period for which snow stays on the ground before melting, there is some evidence to suggest it may be increasing the intensity of snowstorms by adding moisture to the air (Harvey, 2020); (Janoski, 2018). In any case, the challenges of winter are not likely to disappear anytime soon. For vulnerable populations, winter – and deep snow in particular – can greatly compound the difficulties they are already facing.

For children and the able-bodied, blizzards can be a beautiful, even magical experience that momentarily suspends the hectic pace of modern life, covering the urban landscape in a blanket of crystalline white that can be navigated only by sled, snowshoe, or ski. For the elderly and disabled, however, deep snow creates a perilous environment that leaves them homebound and trapped for potentially months. In Albany, as in many other cities, streets are cleared of snow more or less quickly by municipal trucks, but it is the responsibility of property owners to keep their sidewalks clear (an extension of the "drivers first" mentality that privileges the rights of people who can afford cars over those that cannot). In neighborhoods with high rates of vacant properties (as much as 30 percent (Stebbins, 2019)) and absent or neglectful landlords, this translates to a patchwork of cleared, uncleared, and badly cleared sidewalks, rendering them unusable for weeks on end out of the year. Prevented from using sidewalks, I frequently see people risking their lives by walking down streets. For those who have difficulty walking, it can be quite difficult to get out of the way of speeding vehicles on roads made more narrow than usual by piled-up snowbanks.

While municipalities may often have laws requiring people to clear their walks, they are often lightly enforced for fear that they would disproportionately penalize the elderly and disabled for whom snow shoveling is a ridiculously strenuous task resulting in numerous injuries and heart attacks (Haney C. R., 2020). As mechanical snowblowers are expensive tools that require considerable storage space, owning and keeping one is not a realistic option for most urban residents. To make matters worse, municipal plows in their haste to clear the streets will commonly push snow mounds onto (already cleared) sidewalks, creating mountains of ice nearly impossible to cut through (Figure 5.3).

A large part of the issue is that sidewalks are in a gray area in terms of being public or private space. While considered public in that it is illegal to obstruct them, responsibility for maintenance and repair of sidewalks is placed upon the property owners adjacent to them. An in-depth analysis of the problem of snowy sidewalks (Rannila, 2016) concludes that sidewalk-as-commons policies are inherently doomed in a neoliberal city where cars and their drivers are valued over pedestrians. The article argues that attempts by citizens to collectively clear sidewalks play into the interests of "bootstrapping" urban austerity measures and are classic instances of grassroots organizations taking on the responsibilities of government in a "rolling back of the state". While I believe that clearing sidewalks should ultimately be the responsibility of municipalities, and that they should be thought of as public spaces, I am also realistic that this is politically unlikely to happen in rust belt cities that are financially strapped and fearful of "slip and fall" litigation. Could, however, an ethic of the urban commons be applied to the situation? Could

Figure 5.3 Unplowed Sidewalks

Source: Scott Kellogg

citizens organize to cooperatively clear snow from city blocks while simultaneously demanding greater municipal accountability? Collectively purchased snow blowers? A municipally funded citizens snow-corps spraying ice-melting beet sugar (Muthumani, 2017)? The bottom line is that clear and passable sidewalks should not be an exclusive amenity for the rich, especially as it is the poor who rely on them the most as it is their only way to get around.

Conclusion

As climate change advances, global air quality worsens, and atmospheric uncertainty grows, it is imperative that humanity reconfigures its relationship with the underregarded element of air. Of all the aspects of the urban ecosystem discussed in this book: soil, water, air, waste, and biodiversity, air has been the most difficult to write about. Its historical neglect, fluidity, near intangibility, and invisibility make it harder to put boundaries around, fix to a point, describe, or feel familiar with. This chapter also contains little in the way of practical, DIY project based around air. The reason is that I have been honestly hard-pressed to think of many that would feel meaningful and not just "added on". These types of activities that could be used as pedagogical tools for building human–aero connections and appreciation of the aerial commons are unknown to me. What could these resemble? Balloon mapping? Use of drones for aerial photography? Real-time sensors that change color accordingly to the AQI report? I am interested in exploring possibilities for creating familiarity between urban residents and air that are fun, engaging, and emphasize the social and economic aspects of atmospheres and climate. At minimum, I hope that discussion and research will continue on charting "aerographies" (Brown et al., 2020) and the multiple ways that human processes are cocreated with atmospheres and climatic ecologies.

Notes

1 Of course, as I write this in the midst of the 2020 coronavirus pandemic, masks have become symbols of collective responsibility and mutual concern.

2 Based on factors of effectiveness of removing chemicals, ease of cultivation, insect resistance, and transpiration rates.

3 This effect is noticeable by placing your hand on a west-facing brick wall and still feeling heat hours after sunset.

4 While the urban heat island effect is typically discussed in a negative context, it may provide advantages to some. For gardeners, the UHI can extend the growing season considerably. According to one study, this can add up to around a week in both the fall and the spring – about fifteen days in total. The range of affected areas is considerable, with areas up to six miles outside of cities experiencing some related warming. An interesting counterexample to urban heat island cities is Phoenix, Arizona, which is actually cooler in the day than the surrounding nonurban areas. The huge number of pools and irrigated lawns actually produce an evapotranspiration effect not experienced in the surrounding desert lands.

Works cited

Abram, D. (2018). The commonwealth of breath. In *Atmospheres of breathing* (p. 263). Albany, NY: SUNY Press.

Amato, P. (2012). Clouds provide atmospheric oases for microbes. *Microbe*, *7*(3), 119–123.

Avolio, M. L. (2018). Biodiverse cities: The nursery industry, homeowners, and neighborhood differences drive urban tree composition. *Ecological Monographs*, *88*(2), 259–276.

Aydogan, A., & Cerone, R. (2021). Review of the effects of plants on indoor environments. *Indoor and Built Environment*, *30*(4), 442–460.

Barwise, Y. a. (2020). Designing vegetation barriers for urban air pollution abatement: A practical review for appropriate plant species selection. *Climate and Atmospheric Science*, *3*(1), 1–19.

Beacham, A. M. (2019). Vertical farming: A summary of approaches to growing skywards. *The Journal of Horticultural Science and Biotechnology*, *94*(3), 277–283.

Beck, U. (1995). *Ecological enlightenment: Essays on the politics of the risk society*. London: Humanities Press International.

Bo, M. e. (2020). A combined citizen science – modelling approach for NO2 assessment in Torino Urban Agglomeration. *Atmosphere*, *11*(7), 721.

Brito, V. V. (2020). Urban food forestry and its role to increase food security: A Brazilian overview and its potentialities. *Urban Forestry & Urban Greening*. USDA report. Retrieved from https://pubag.nal.usda.gov/catalog/7095314

Brown, N., Buse, C., Lewis, A., Martin, D., & Nettleton, S. (2020). Air care: An "aerography" of breath, buildings and bugs in the cystic fibrosis clinic. *Sociology of Health & Illness*, *42*(5), 972–986.

Bunge, A. e. (2019). Urban foraging for food security and sovereignty: Quantifying edible forest yield in Syracuse, New York using four common fruit-and nut-producing street tree species. *Journal of Urban Ecology*, *5*(1).

Castro, J. e. (2018). Edible urban forests as part of inclusive, sustainable cities. *Unasylva*, 69(250), 59–65.

Chay, K. Y. (2003). The impact of air pollution on infant mortality: Evidence from geographic variation in pollution shocks induced by a recession. *The Quarterly Journal of Economics*, *118*(3), 1121–1167.

Chen, L.-J. e. (2017). An open framework for participatory PM 2.5 monitoring in smart cities. *IEEE Access*, 5, 14441–14454.

Chilson, P. B. (2017). *Aeroecology*. Cham: Springer International Publishing.

Clark, K. H. (2013). Introducing urban food forestry: A multifunctional approach to increase food security and provide ecosystem services. *Landscape Ecology*, *28*(9), 1649–1669.

Dayaneni, G. (2009). Carbon fundamentalism vs. climate justice. *Race, Poverty & the Environment*, *16*(2), 7–11.

Dutta, P. e. (2009). Common sense: Participatory urban sensing using a network of handheld air quality monitors. *Proceedings of the 7th ACM Conference on Embedded Networked Sensor Systems* (pp. 349–350). Retrieved from https://doi.org/10.1145/1644038.1644095.

English, P. B. (2017). The imperial county community air monitoring network: A model for community-based environmental monitoring for public health action. *Environmental Health Perspectives*, *125*(7).

EPA. (2020a). *Heat island effect*. United States Environmental Protection Agency. Retrieved from www.epa.gov/heatislands

EPA. (2020b). *Particulate Matter (PM) pollution*. United States Environmental Protection Agency. Retrieved from www.epa.gov/pm-pollution/particulate-matter-pm-basics

Erell, E. D. (2012). *Urban microclimate: Designing the spaces between buildings*. London: Routledge.

Fares, S. e. (2017). Carbon sequestration by urban trees. In *The urban forest* (pp. 31–39). Cham: Springer.

Field, J. J. (2020). *Urban tree canopy governance and redlined neighborhoods: An analysis of five cities*. Dissertation, Massachusetts Institute of Technology.

Fischer, L. K. (2020). Connecting people to biodiversity in cities of tomorrow: Is urban foraging a powerful tool? *Ecological Indicators, 112*.

Freytas-Tamura, K. d. (2017, February 17). A push for diesel leaves London gasping amid record pollution. *The New York Times*.

Gaither, C. J., Aragón, A., Madden, M., Alford, S., Wynn, A., & Emery, M. (2020). "Black Folks Do Forage": Examining wild food gathering in Southeast Atlanta communities. *Urban Forestry & Urban Greening*, 56, 126860.

Garrison, J. D. (2018). Environmental justice in theory and practice: Measuring the equity outcomes of Los Angeles and New York's "Million Trees" campaigns. *Journal of Planning Education and Research*. Retrieved from https://doi.org/10.1177/0739456X18772072.

Ghertner, D. A. (2020). Airpocalypse: Distributions of life amidst Delhi's polluted airs. *Public Culture, 32*(1), 133–162.

Golumbic, Y. N.-T. (2019). User centered design of a citizen science air-quality monitoring project. *International Journal of Science Education, Part B*, 9(3), 195–213.

Gori, A. F. (2019). Reprint of: Growing healthy food under heavy metal pollution load: Overview and major challenges of tree based edible landscapes. *Urban Forestry & Urban Greening, 45*.

Górska, M. (2016). *Breathing matters: Feminist intersectional politics of vulnerability*. Linkoping, Switzerland: Linkoping University.

Graham, S. (2015). Life support: The political ecology of urban air. *City, 19*(2–3), 192–215.

Hagler, G. S. (2012). Field investigation of roadside vegetative and structural barrier impact on near-road ultrafine particle concentrations under a variety of wind conditions. *Science of the Total Environment, 419*, 7–15.

Hamilton, W. D. (1998). Spora and Gaia: How microbes fly with their clouds. *Ethology Ecology & Evolution, 10*(1), 1–16.

Haney, C. (2020). Beyond "Snow Shoveler's Infarction": Broadening perspectives on winter health risks. *Geography Compass, 14*(7).

Haney, C. R. (2020). *Winter weather hazards: Injuries and fatalities associated with snow removal*. Dissertation, Mississippi State University.

Harlan, S. L. (2007). In the shade of affluence: The inequitable distribution of the urban heat island. *Research in Social Problems and Public Policy, 15*, 173–202.

Harvey, C. (2020, January). Love snow? Here's how it's changing. *Scientific American*.

Hawken, P. e. (2017). *Drawdown: The most comprehensive plan ever proposed to reverse global warming*. New York: Penguin.

Helden, A. J. (2012). Urban biodiversity: Comparison of insect assemblages on native and non-native trees. *Urban Ecosystems, 15*(3), 611–624.

Houston, A. A. (2015). *Real time data*. Retrieved from www.airalliancehouston.org/air-quality-now/

Irga, P. J. (2018). The phytoremediation of indoor air pollution: A review on the technology development from the potted plant through to functional green wall biofilters. *Reviews in Environmental Science and Bio/Technology, 17*(2), 395–415.

Jacke, D. a. (2005). *Edible forest gardens, Volume II: Ecological design and practice for temperate-climate permaculture*. Hartford, VT: Chelsea Green Publishing.

Janoski, T. P. (2018). Effects of climate change on wind-driven heavy-snowfall events over eastern North America. *Journal of Climate, 31*(22), 9037–9054.

Kaufman, A. e. (2017). A citizen science and government collaboration: Developing tools to facilitate community air monitoring. *Environmental Justice, 10*(2), 51–61.

Kendal, D. C. (2014). Global patterns of diversity in the urban forest: Is there evidence to support the 10/20/30 rule? *Urban Forestry & Urban Greening, 13*(3), 411–417.

Kim, K. J. (2010). Variation in formaldehyde removal efficiency among indoor plant species. *HortScience, 45*(10), 1489–1495.

Klinenberg, E. (2015). *Heat wave: A social autopsy of disaster in Chicago*. Chicago, IL: University of Chicago Press.

Koo, B. W., Boyd, N., Botchwey, N., & Guhathakurta, S. (2019). Environmental equity and spatiotemporal patterns of urban tree canopy in Atlanta. *Journal of Planning Education and Research*. Retrieved from https://doi.org/10.1177/0739456X19864149.

Kull, C. A. (2018). Critical invasion science: Weeds, pests, and aliens. In *The Palgrave handbook of critical physical geography* (pp. 249–272). Cham: Palgrave Macmillan.

Kumar, P. e. (2014). Ultrafine particles in cities. *Environment International*, 66, 1–10.

Kumar, P. e. (2015). New directions: Air pollution challenges for developing megacities like Delhi. *Atmospheric Environment, 122*, 657–661.

Landor-Yamagata, J. L. (2018). Urban foraging in Berlin: People, plants and practices within the metropolitan green infrastructure. *Sustainability, 10*(6).

Locke, D. H., Hall, B., Grove, J. M., Pickett, S. T., Ogden, L. A., Aoki, C., . . . & O'Neil-Dunne, J. P. (2021). Residential housing segregation and urban tree canopy in 37 US Cities. *npj Urban Sustainability, 1*(1), 1–9.

Lohr, V. I. (2014). Urban trees worldwide have low species and genetic diversity, posing high risks of tree loss as stresses from climate change increase. *XXIX International Horticultural Congress on Horticulture: Sustaining Lives, Livs, Livelihoods and Landscapes (IHC2014)*, V 1108.

Maas, R., & Grennfelt, P. (2016). Towards cleaner Air-CLRTAP scientific assessment report 2016. In *EMEP-Steering body and Working Group on effects-convention on long-range transboundary air pollution* (Vol. 1). United Nations.

MacIvor, J. S. (2016). Air temperature cooling by extensive green roofs in Toronto Canada. *Ecological Engineering*, 95, 36–42.

Mahajan, S., & Kumar, P. (2020). Evaluation of low-cost sensors for quantitative personal exposure monitoring. *Sustainable Cities and Society*, 57, 102076.

Mahajan, S. e. (2020). A citizen science approach for enhancing public understanding of air pollution. *Sustainable Cities and Society*, 52.

Manes, F. e. (2012). Urban ecosystem services: Tree diversity and stability of tropospheric ozone removal. *Ecological Applications, 22*(1), 349–360.

Manigrasso, M. e. (2017). Pedestrians in traffic environments: Ultrafine particle respiratory doses. *International Journal of Environmental Research and Public Health, 14*(3), 288.

McLain, R. e. (2012). Producing edible landscapes in Seattle's urban forest. *Urban Forestry & Urban Greening, 11*(2), 187–194.

McLain, R. J. (2014). Gathering "wild" food in the city: Rethinking the role of foraging in urban ecosystem planning and management. *Local Environment, 19*(2), 220–240.

Morello-Frosch, R. M. (2001). Environmental justice and southern California's "Riskscape": The distribution of air toxics exposures and health risks among diverse communities. *Urban Affairs Review, 36*(4), 551–555.

Murphy, M. W. (2013). Formaldehyde levels in FEMA-supplied travel trailers, park models, and mobile homes in Louisiana and Mississippi. *Indoor Air, 23*(2), 134–141.

Muthumani, A. a. (2017). Effectiveness of liquid agricultural by-products and solid complex chlorides for snow and ice control. *Journal of Cold Regions Engineering, 31*(1).

Negi, R. (2020). Urban air. *Comparative Studies of South Asia, Africa and the Middle East, 40*(1), 17–23.

Nieuwenhuis, M. (2013). Terror in the air in Istanbul. *Society and Space: open site*. Retrieved October 5, 2016, from www.societyandspace. com/2013/06/20/marijnnieuwenhuis-terror-in-the-air-in-istanbul/# comments

Nieuwenhuis, M. (2018). Atmospheric governance: Gassing as law for the protection and killing of life. *Environment and Planning D: Society and Space*, *36*(1), 78–95.

Novotny, E. V. (2008). Increase of urban lake salinity by road deicing salt. *Science of the Total Environment*, *406*(1–2), 131–144.

Nyelele, C. a. (2020). Urban forestry & urban greening. *The Equity of Urban Forest Ecosystem Services and Benefits in the Bronx, NY.*, *53*.

O'Dowd, C. D. (2002). Marine aerosol formation from biogenic iodine emissions. *Nature*, *417*(6889), 632–636.

Ogden, L. E. (2014). Life in the clouds. *BioScience*, *64*(10), 861–867.

Ottinger, G. (2010). Buckets of resistance: Standards and the effectiveness of citizen science. *Science, Technology, & Human Values*, *35*(2), 244–270.

Ottosen, T.-B. a. (2020). The influence of the vegetation cycle on the mitigation of air pollution by a deciduous roadside hedge. *Sustainable Cities and Society*, *53*.

Pal, J. S. (2015). Future temperature in Southwest Asia projected to exceed a threshold for human adaptability. *Nature Climate Change*, *6*(2), 197.

Park, H. e. (2019). Urban food systems that involve trees in Northern America and Europe: A scoping review. *Urban Forestry & Urban Greening*, *45*.

Poe, M. R. (2014). Urban foraging and the relational ecologies of belonging. *Social & Cultural Geography*, *15*(8), 901–919.

Ragnedda, M. (2017). *The third digital divide: A Weberian approach to digital inequalities*. London: Taylor & Francis.

Rannila, P. a. (2016). Syracuse, sidewalks, and snow: The slippery realities of public space. *Urban Geography*, *37*(7), 1070–1090.

Riolo, F. (2019). The social and environmental value of public urban food forests: The case study of the Picasso Food Forest in Parma, Italy. *Urban Forestry & Urban Greening*, *45*.

Rizwan, A. M. (2008). A review on the generation, determination and mitigation of Urban Heat Island. *Journal of Environmental Sciences*, *20*(1), 120–128.

Robitu, M. e. (2006). Modeling the influence of vegetation and water pond on urban microclimate. *Solar Energy*, *80*(4), 435–447.

Sanchez, L. a. (2019). Cooling Detroit: A socio-spatial analysis of equity in green roofs as an urban heat island mitigation strategy. *Urban Forestry & Urban Greening*, *44*.

Shackleton, C. M. (2017). Urban foraging: A ubiquitous human practice overlooked by urban planners, policy, and research. *Sustainability*, *9*(10).

Sheil, D. (2018). Forests, atmospheric water and an uncertain future: The new biology of the global water cycle. *Forest Ecosystems*, *5*(1), 1–22.

Shendell, D. G. (2007). The Outdoor Air Quality Flag Program in Central California: A school-based educational intervention to potentially help reduce children's exposure to environmental asthma triggers. *Journal of Environmental Health*, *70*.

Sloterdijk, P. (2009). *Terror from the Air*. Semiotext. Cambridge, MA: MIT Press.

Sobel, D. (2004). *Place-based education: Connecting classrooms & communities*. Great Barrington, MA: Orion Society.

Stebbins, S. (2019, February 26). Ghost towns: 30 Zip codes with the highest vacancy rates in the US. *USA Today*.

Steele, W. L. (2015). Urban climate justice: Creating sustainable pathways for humans and other species. *Current Opinion in Environmental Sustainability*, *14*, 121–126.

Steffens, J. T. (2012). Exploration of effects of a vegetation barrier on particle size distributions in a near-road environment. *Atmospheric Environment*, *50*, 120–128.

Takane, Y. e. (2019). Urban warming and future air-conditioning use in an Asian megacity: Importance of positive feedback. *Climate and Atmospheric Science*, *2*(1), 1–11.

Tam, N. F. (1987). Heavy metal pollution in roadside urban parks and gardens in Hong Kong. *Science of the Total Environment*, 59, 325–328.

Tellström, S. (2014). *Urban agroforestry: For developing ecosystem services in urban forests*. Thesis. Mid Sweden University, Faculty of Science, Technology and Media, Department of Ecotechnology and Sustainable Building Engineering.

Threlfall, C. G. (2017). Increasing biodiversity in urban green spaces through simple vegetation interventions. *Journal of Applied Ecology*, *54*(6), 1874–1883.

Tsilini, V. e. (2015). Urban gardens as a solution to energy poverty and urban heat island. *Sustainable Cities and Society*, *14*, 323–333.

Waring, M. S. (2016). Bio-walls and indoor houseplants: Facts and fictions. *Microbiomes of the Built Environment: From Research to Application, Meeting, 3*.

Weerakkody, U. e. (2018). Evaluating the impact of individual leaf traits on atmospheric particulate matter accumulation using natural and synthetic leaves. *Urban Forestry & Urban Greening*, *30*, 98–107.

Wightman, K. E. (1999). *Good tree nursery practices: Practical guidelines for community nurseries*. Nairobi, Kenya: International Centre for Research in Agroforestry.

Wolverton, B. (2010). *Plant-based air filters for formaldehyde remediation in FEMA trailers*. Picayune, MS: Wolverton Environmental Services, Inc.

Wolverton, B. C. (1997). *How to grow fresh air: 50 Houseplants to purify your home or office*. New York: Penguin Books.

World Health Organization. (2016). Ambient air pollution: A global assessment of exposure and burden of disease. *Clean Air Journal*, *26*(2), 6.

Yang, D. S. (2009). Screening indoor plants for volatile organic pollutant removal efficiency. *HortScience*, *44*(5), 1377–1381.

Yang, J. Y. (2015). Ranking the suitability of common urban tree species for controlling PM2.5 pollution. *Atmospheric Pollution Research*, 6(2), 267–277.

6 Urban compost justice

In my work as an urban socio-ecological educator, one of the most important concepts that I try to get across to my audiences is the critical difference between sustainable and non-sustainable agricultural societies. Although it's admittedly an oversimplification, this difference can be broken down to saying, "sustainable societies return their wastes to the soil, non-sustainable societies do not". History provides many examples of cultures growing food on the same land for generations simply by carrying out the practice of returning their wastes to the soil (King, 2011). When nutrients are retuned to soil in the form of composted food wastes and manures, soil microorganisms thrive and maintain soil tilth. In contrast, those cultures that grow crops on the same land year after year without putting anything back will eventually deplete their soils, leaving them vulnerable to erosion and ultimately, desertification. These degenerative growing practices then become ingrained in culture and repeated as expansionist societies invade new territories following the destruction of their homelands (Montgomery, 2012).

Western capitalist society, with its high-consumption linear throughput economy, clearly functions like the latter example. Industrial-scale food production on land, made possible through intensive applications of petroleum-derived chemical fertilizers, biocides, and energy, results in the depletion and destruction of soils. The food grown on these lands is then transported intercontinentally where much of it goes to waste, ultimately ending up in landfills where it degrades into methane, a potent greenhouse gas. This process is a global-scale manifestation of "metabolic rift", a phenomenon described in depth by political ecologists (Swyngedouw, 2006) that results in the degradation of both land and people. Western cities, with their food production externalized to their hinterlands and their populations largely ignorant of agroecological processes, are poorly attuned to the feedback of the fertility of surrounding soils. This alienation and insensitivity perpetuates the metabolic rift, turning cities into soil nutrient vacuums – pulling in resources and exporting toxicity, poisoning the lands and people both within and outside the city limits (McClintock, 2012).

Urban ecosystem justice seeks to repair this broken cycle so that the vitality of soils, both urban and rural, may be restored. Composting – the process of microbes transforming organic waste into soil – holds the key for repairing this broken link and for promoting both social and ecological regeneration. By intercepting and

DOI: 10.4324/9780367858711-7

composting the enormous volumes of organic matter currently being thrown into landfills each day, soil fertility can be restored, toxins remediated, economic opportunities created, food security promoted, and climate change resisted.

The discussion of all forms of urban waste is vast, multifaceted, and critically important; however, for the purposes of this book I will focus my analysis on the composting of urban organic wastes, and specifically food residues, believing it to have the greatest relevance to urban ecosystem justice.[1] Again, this has to do with the centrality and importance of urban soils to the framework and potential for citizen control over the process. Apart from gaining access to land, poor soil quality is one of the biggest limiting factors of food production in the city. Urban soils are typically nonexistent (i.e., paved over), severely degraded (nutrient poor, thin, compressed), or contaminated. The remedy for any of these soil maladies is to apply finished compost. Compost can create soil where it is nonexistent, improve and revitalize degraded and dead soils, and can facilitate the degradation or reduce the biological availability of toxins in contaminated soil (Attanayake, 2015). In this regard, the solution to restoring the health of urban soils exists within cities themselves in the form of food wastes and carbon residues. While it is paradoxical that this surplus of food waste only exists as a by-product of our wastefully extravagant economy, it is nonetheless essential to utilize. By converting food wastes into soils and in turn growing nutrient-dense foods within city limits, it becomes possible to address systemic issues related to food access and justice. Although it is by no means a total fix to what are genuinely huge, wicked, and systemic issues, it can be a key component among a multiplicity of interventions.

Waste in the urban metabolism

While this book already dedicates an entire chapter to the issue of urban soils, the matter of composting and waste is both unique and significant enough to deserve its own section. "Waste", in the context of this book, is somewhat of an outlying concept. It does not represent a "biophysical sphere" in the same sense that soil, water, and air do, each having clear boundaries defined by their physical material states. Rather, waste is a phase of metabolic processes – it can be treated either as potentially valuable resource or as a toxic externality.

In order to better understand waste processes, it can be helpful to view them through the lens of urban metabolic theory.

Linear waste

In a linear metabolic system, waste is equal to output, discharge, emissions – unwanted and toxic materials (trash, wastewater, particulates, CO_2, heat) dumped into the atmosphere, sewers, incinerators, and landfills. While some resource recovery is possible from linear metabolic waste, it is done only with considerable energetic expense and usually results in a recovered material with high toxic concentrations (biosolids from sewage treatment and landfill gas, for example). Furthermore, as thoroughly documented by environmental justice scholars

(Agyeman, 2005), it is commonly lower-income and marginalized populations who bear the greatest risk of exposure to these toxic and noxious outputs on account of their proximity to their places of storage/processing/discharge. Increasingly, the flow paths for metabolic waste outputs are growing longer with toxic burdens being externalized to communities living far from their points of production. This has long been the case for airborne pollutants, which can be transported long distances through atmospheric "grasshopper" effects (Gouin, 2004) but more so now for solid waste which is being shipped greater distances as more and more peri-urban landfills reach capacity (Hoornweg, 2013).

Globally, as much as one-third of the food grown for human consumption goes to waste on account of a variety of factors at every level from harvest, storage, transportation, and market (Gustavsson, 2015). Not only is this a waste of the food itself but also of the resources used and greenhouse gas emissions resulting in its production. While food spoilage occurs in both the Global North and the South, the majority occurs in the former, with nearly 40 percent of food being wasted in the United States (Gunders, 2012). Conventionally, food waste is put into landfills along with other types of organic matter including wood chips, grass clippings, brown leaves, paper wastes, etc. Organic matter placed into landfills takes up a considerable amount of space in the fill, problematic as society nears "peak waste" (Hoornweg, 2013). Upon entering the landfill, this relatively uncontaminated organic matter is mixed with other forms of waste, buried below ground, and compacted with heavy machinery. The compaction process squeezes out most of the oxygen in the landfill, causing the organic matter to be decomposed under anaerobic conditions by methanogenic bacteria. The methanogens produce methane, a greenhouse gas nearly twenty-five times more potent than carbon dioxide, as a by-product (EPA, 2017). Fugitive methane escaping from landfills alone contributes to between 3 percent and 12 percent of greenhouse gas emissions in the United States and accounts for 24 percent of total methane emissions in the United States. While some landfills are equipped with methane recovery systems or Landfill Gas-to-Energy (LFGTE) facilities, these usually capture only a percentage of the escaping gas, as little as 20 percent, according to the IPCC (Intergovernmental Panel on Climate Change). While having LFGTE systems are arguably preferable to having no gas recovery systems in existing landfills, their use should not be used as a justification for continuing the practice of putting organics in the landfill, as nearly any other method of handling organic wastes will result in reduced greenhouse gas production. To increase short-term methane production, LFGTE operators usually create "wet cells" as opposed to "dry tombs" in landfills by soaking them with water. This practice boosts gas yields but results in an increase of the total amount of fugitive methane emissions. Lastly, landfill gas can be highly toxic as it is mixed with a variety of volatile pollutants in the landfill. Regulating these toxic emissions is difficult as there are insufficient procedures for measuring non-point and diffuse emissions over a landfill's surface (Sierra, 2010). In conclusion, the process of landfilling organics thus described is highly degenerative not only for taking up space in landfills and for producing gas emissions but also for ensuring the impossibility of composting the disposed

Figure 6.1 Linear Food Waste
Source: Scott Kellogg

organic matter into soil (Figure 6.1). Unfortunately, the mindset of thinking of organic waste as fuel for LFGTE systems and then of recovered gas as a revenue stream is pervasive. A high-ranking city official explained to me at one time in a personal communication their philosophical opposition to composting and belief that organics "should go into the landfill, as they then turned into methane which then made the city money".

Cyclical waste systems

In an urban ecosystem justice context, it will be necessary to completely rethink the idea of waste. Guidance for how to do so can be gained through observance of cyclical ecosystems and regenerative pre-industrial human societies. In cyclical ecosystems, "waste" means something fundamentally different than it does in a system with linear throughputs. Rather than falling at one end of the input–output dichotomy, cyclical waste represents the by-products produced by organisms, that is, oxygen, carbon dioxide, sweat, urine, feces, dead leaves, corpses, etc. These are rapidly taken up by other living things in the ecosystem, broken down into minerals that reenter the planet's biogeochemical cycles, or buried in the earth's geological layers. In terrestrial ecosystems, much of this material ultimately ends up becoming incorporated into in the planet's organic soil horizon – the complex interface of minerals, organic molecules, moisture, air, and living organisms that create the basis of land-based food chains. As soils increase in depth, so does the fertility and overall productivity of an ecosystem, along with its capacity to retain moisture, buffer climate changes, filter water, sequester carbon, support biodiversity, resist erosion, etc. In this regard, "wastes" are not undesirable materials to be rapidly expelled but rather resources to be used for building greater ecological resilience. Regenerative human societies that practiced sustainable agriculture in the same lands for prolonged periods understood this principle and incorporated that knowledge into their cultures. For instance, ancient Amazonian cultures were able to build highly fertile soils through the organic wastes, charcoal, and pottery shards discarded in their middens (Balée, 2006).

If cities are to function more like cyclical ecosystems, waste must be treated like a valuable resource that can be used to increase ecosystem health rather than a negative industrial externality. Like the other "spheres" of the urban ecosystem, toxicity in the waste stream presents a significant barrier to achieving urban ecosystem justice. A critical part of successfully interrupting and hijacking the linear throughput urban metabolisms is to prevent the mixing of toxic materials into uncontaminated materials to begin with. This is particularly important in the case of consumer products and building materials whose infusion with toxic substances prevents them from being safely composted or otherwise reused (pressure-treated wood, which is imbued with toxic metals, is a good example of this). The disposal of these toxic materials in landfills and incinerators results in the production of considerable toxic emissions and residue that affect the health of communities near and far. In consideration of this, it is critically important to advance green chemistry technologies to replace toxic consumer products with materials that are ultimately nontoxic and compostable (Anastas, 2000).

Furthermore, movement must be made away from municipal combined sewer systems. Blending freshwater with sewage, stormwater runoff, industrial wastes, pharmaceuticals, and food wastes, sewers take relatively uncontaminated materials (water, food wastes, sewage) and mix them with toxic materials and street runoff, resulting in a contaminated end product that can poison lands when applied to them (Protection, 2019).

Basics of composting

Before proceeding with a discussion about the political and social dimensions of composting, it is important to be acquainted with the basic science of the composting process. Beautiful in both its simplicity and complexity, familiarity with composting will lend to appreciating its potential for community empowerment.

A compost pile is essentially a microbial bioreactor. When organic wastes are mixed to achieve the correct ratio of carbonaceous (carbon-based) to nitrogenous (nitrogen-based) materials in a pile of at minimum of a cubic yard, and when sufficient oxygen and moisture are present, the composting process begins. Indigenous bacteria will begin consuming the wastes, generating heat in the process that accelerates the rate of decomposition. This period of active decomposition can last anywhere between three weeks to several months, depending on the availability of oxygen, moisture, nutrients, etc. During this period a nearly 50 percent reduction in the volume of the compost pile occurs, with much of its mass being lost to an aqueous leachate or escaping in the form of carbon dioxide. Following the active phase and once the temperature of the pile drops, other organisms become active in the pile including fungi, actinomycetes, nematodes, worms, micro-invertebrates, etc. This multispecies sympoietic ecosystem gives the compost a final polish, resulting in a finished product that is dark, moist, crumbly, pH neutral, and smells sweetly of earth, an example of the "Power of the Humusites for a Habitable Multispecies Muddle" (Haraway, 2016). Teeming with nutrients and microbial life, the finished compost can then be used as an amendment to garden soils (Campbell, 1998).

The key to successful composting is to achieve the aforementioned proper balance of carbon- and nitrogen-based materials, respectively also referred to as "browns" and "greens". "Browns" tend to be materials that are dry and woody, while "greens" are more often wet and more recently living. The following table (Table 6.1) orders a few common materials into either category.

The "golden ratio" of carbon to nitrogen necessary for composting is roughly thirty parts carbon to one part nitrogen (30:1). Roughly, this translates to always erring on the side of adding more carbon-based than nitrogen-based materials to compost. A pile that is too heavy in carbon will only be slow to break down while a pile too rich in nitrogen will begin to stink and produce methane (which is of course, contrary to the goals of composting). Regularly composted materials have known relative percentages of carbon to nitrogen on a dry weight basis, and in this regard composting can be managed quite precisely. For most people, however, exact values of carbon to nitrogen are not calculated while building compost piles, and the process becomes more of an intuitive art than a reductive science. It is possible by just relying upon one's observational senses, notably smell, to know that a compost pile has been correctly built and whether any corrections are needed. A simple rule to follow is, "if it stinks, add more carbon". High temperatures are an indicator of a successfully built compost pile (Figure 6.2). When properly constructed, a compost pile may reach a temperature of 160 degrees Fahrenheit (above that temperature bacteria cannot survive and negative feedback loops prevent runaway reactions).

Obtaining an adequately large and dependable supply of carbon is a key challenge in community-based composting. As mentioned, insufficient carbon will result in odors that will offend neighbors and create negative perceptions of compost and composting, particularly if they are unfamiliar with agricultural processes. For this reason, it is critical to have an ample supply of "browns" on site at

Table 6.1 Table of Carbon (Browns) and Nitrogen (Greens)

Carbon/Browns	*Nitrogen/Greens*
Wood chips/sawdust	Food wastes
Straw/old hay	Animal manures
Dead leaves	Grass clippings/weeds
Shredded paper/cardboard	Blood/hair/urine

Figure 6.2 Composting Thermometer
Source: photo by Hannah Katz

all times. My personal favorite form of carbon is the wood chip. Wood chips are carbon-dense relative to their total volume, which makes them space efficient. When piled on top of one another they leave ample space for the infiltration of oxygen, essential for maintaining aerobic conditions in a compost pile. Lastly, they can be obtained in large quantities, typically at no cost. We have accomplished this at Radix by approaching both private and municipal arborists with tow-behind chipper/shredders and offering them a free location to dispose of their chips. They are often pleased to do so if the location is closer and more accessible than where they might otherwise need to be taken, and if they are not contractually bound to deliver chips elsewhere (such as centralized biomass plants). Urban arborists are frequently removing tree limbs and entire trees as keeping trees clear of power lines requires nearly constant maintenance. The politics and ethics of this aside, it ensures a regular supply of carbon to community composters.

Critical composting

Compost and composting have been studied extensively as a biophysical process by agricultural scientists, waste management industries, and waste policymakers. The process has been thoroughly reductively analyzed and can be carried out with a high level of technical precision. While the scientific literature related to composting is robust, there is a shortage of studies regarding the social, power, and political dimensions of composting. Questions such as "who composts?", "who controls access to food wastes?", and "what purposes are composted food wastes used for?" are seldom asked. Do composted food scraps get used to regenerate the health of soils in environmental justice neighborhoods or are they being used to generate profits for a handful of waste handling corporations? Are decentralized and community-scale composting strategies regarded with the same seriousness as centralized and macro-scale programs? What political influence do large-scale composting businesses have that might be used to shut out or illegalize community-scale composting? Could the "right to compost" be considered alongside of the other goals of the "right to the city" movement (Lefebvre, 1996)? It is important to ask these questions in order to ensure that organics recycling and composting remains a process of the commons – one that is owned and controlled by a diverse coalition of stakeholders including grassroots activists, local communities, and small mission driven nonprofits. While not immediately pressing, a great potential future threat to community-based composting lies in the growth of corporate composters for whom organic resources are solely a commodity – a form of refuse no different from bottles, cans, and trash. These entities stand to profit from the long-distance hauling of organic wastes, intensive industrial scale composting, and sales of combustible gas generated by centralized anaerobic digesters. Separated from and indifferent to agricultural needs, industrial-scale composting produces an inferior (and frequently contaminated) end product. None of the profits or resources they generate will ever be circulated back to environmental justice communities where they are greatly needed.

While industrial composting is without a question immensely preferable to incineration or landfilling, small and community-scale composting is essential for attaining urban ecosystem justice. Proactive and preventative work must be done to ensure that composting remains a process of the commons: localized, decentralized, uncontaminated, and community controlled. Complimentarily, there is a need for the development of "critical composting studies" that analyze the social, governance, and economic aspects of how organic wastes are processed and what communities benefit from them.

Already there are marked differences in opinion about how organics recycling should proceed, with governmental agencies typically preferring concentrated, centralized, and technologically sophisticated methods over decentralized, community-based, and small-scale approaches chosen by grassroots organizations. To illustrate this difference, and to assist in framing the core issues of a critical composting analysis, I will examine two divergent versions of food waste recovery hierarchies.

Source reduction

It should be mentioned that much of the food being thrown into landfills is still perfectly fit for human consumption. Large quantities of edible food are regularly disposed of simply for reasons of convenience to the food industry – supermarkets must routinely be clearing their shelves of older produce to make space for new inventory as there is little or no market for its resale and no means to accordingly and finely adjust the shipment of food on the production end. Considering this, it is most important for food that is still edible to be put into the hands of hungry people. The efforts of food recovery organizations and food pantries are critical to achieving the collection and distribution of this otherwise squandered resource. This view is supported by the federal government's "food diversion" pyramids that propose a hierarchy of preferred uses for unsellable food (Papargyropoulou, 2014). In ideal circumstances, relationships can be established between food pantries and composters where foods that are unwanted or unfit for consumption can be composted following efforts to distribute them. The Radix Center maintains such a relationship with the Capital City Rescue Mission, a food pantry located merely a block away. The mission collects food from large supermarkets in the region in addition to receiving donations from farms and caterers. This collected food is then made available for free to city residents three days each week. Following the end of the pantry, Radix collects the unwanted and spoiled food in wheelbarrows which is then composted on the farm. Bringing students to observe this process firsthand provides a valuable educational opportunity. Not only do they get to see what food waste diversion pyramid strategies look like in action, but they are made aware of the enormous quantity of food that typically otherwise goes to waste. Seeing this firsthand opens the conversation about food systems on a global scale and leads to the question of whether food insecurity is a problem of either limited production or inadequate distribution.

Food waste recovery pyramid comparison

Food waste recovery hierarchies are graphic diagrams designed to assist with decision-making regarding preferred uses for food wastes. The first depicted is the "Hierarchy for Reducing and Recycling Food Scraps and Other Organic Discards" as created by the Institute for Local Self Reliance (ILSR), a nonprofit organization that advocates for community control of resources (Figure 6.3). The second is the "Food Recovery Hierarchy", as developed by the United States Environmental Protection Agency (EPA) (Figure 6.4).

Both schemas prioritize source reduction and edible food rescue as the first and second most preferred uses of food wastes. Additionally, they also both list landfilling and incineration as the least desirable option for food waste management. The middle sections of both diagrams show notable differences, however, they are worth analyzing. There is an inversion of preferences in the two charts for composting and industrial applications, with EPA prioritizing industrial uses and ILSR advocating primarily for composting. ILSR goes so far as to break the "composting" category down into three subcategories: residential/backyard, small-scale

Figure 6.3 Composting Demonstration

Source: Scott Kellogg

decentralized, and centralized composting/anaerobic digestion. Their respective order on the chart is reflective of ILSR's advocacy for citizen and community-controlled approaches to composting being given greater consideration over centralized and industrial approaches (the "feed animals" category present on the EPA diagram (Figure 6.5) is not included on the ILSR pyramid – likely as it is regarded as a form of composting and could accordingly operate on residential, small, and centralized scales).

EPA, conversely, shows a strong preference for centralized and industrial uses for food wastes. The EPA defines "industrial uses" as anaerobic digestion along with fat, oils, and grease (FOG) rendering. The centralized, concentrated, and technologically sophisticated methods and needs of industry and mega-scale agribusiness clearly are prioritized over any form of composting (regardless of scale) by the EPA, which comes in fifth place in their order of preferred uses. Fundamentally, these charts represent two dominant ways of thinking about food scraps reuse: they can either be used to enrich corporations and industrial agri-businesses or be used to foster greater community self-reliance and control while promoting ecological regeneration.

In the next section, I will explain centralized anaerobic digesters/digestion (AD) in greater depth and contrast it against smaller-scale community-based composting strategies.

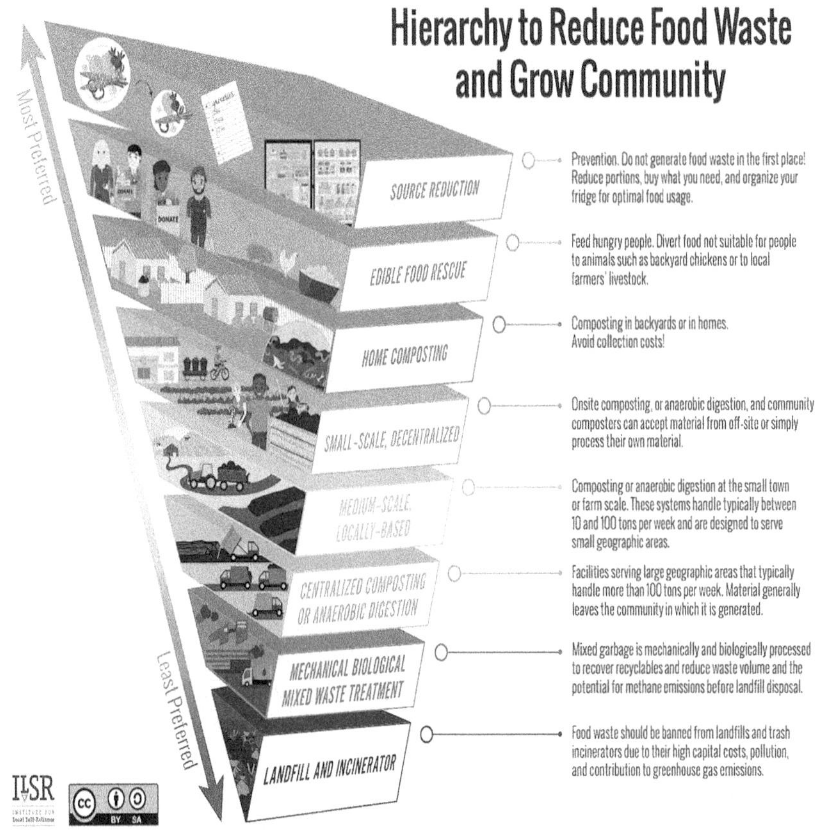

Figure 6.4 Hierarchy for Reducing Food Scraps and Other Organic Discards – Institute for Local Self Reliance

Source: https://ilsr.org/food-waste-hierarchy/. The figure comes from the **Institute for Local Self-Reliance (www.ilsr.org)**, a national nonprofit organization working to strengthen local economies and redirect waste into local recycling, composting, and reuse industries. It is reprinted here with permission.

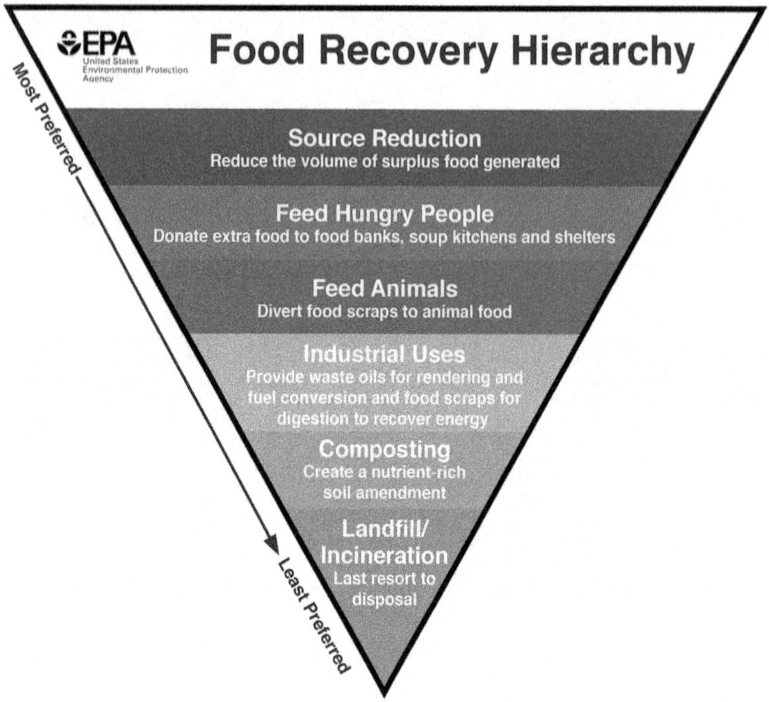

Figure 6.5 Food Recovery Hierarchy – US EPA

Community composting

On account of the confluence of a variety of ecologically and economically limiting factors along with considerable social pressures, a handful of urban municipalities have taken the steps to begin city food waste collection and composting programs. A considerable range of approaches have been taken to address the need for municipal composting. These have included large waste handling corporations such as Waste Management expanding the range of their services, mid-range haulers bringing food scraps to regional composting facilities, and "micro-haulers" or community-based composting initiatives working primarily within city limits on small parcels of land (Schaffer, 2016).

Despite the importance of composting, however, many cities (particularly smaller and less affluent municipalities) have been slow to act in implementing city-operated programs. In response to the inaction of municipal governments, and from a desire to prevent large corporations from dominating the composting industry, a number of community-based, citizen-sponsored compost collection initiatives have begun. Collecting food wastes from both residents and businesses, these groups are utilizing vacant lots and community gardens to transform food wastes into soil (Hermann et al., 2016).

Community-based composting has many advantages over larger-scale municipal compost collection and mid-range haulers. Perhaps, the most significant of these is the amount of energy and greenhouse gases that are spared by handling the collection and processing of compost locally. Transporting food wastes long distances increases the "food miles" travelled by foods, a metric used to measure the benefits of eating locally (Weber, 2008). When food waste is collected and then transported many miles out of a city to a large commercial composting facility, commonly with diesel-powered trucks, enormous amounts of carbon-based fuels are combusted. Depending on the total distance driven, this energy consumption could negate the reduction in greenhouse gases achieved by separating organics in the first place. If micro-hauling operations move food wastes with bicycle carts or electric or biofueled vehicles, it may be possible to even further reduce greenhouse gas emissions associated with composting. Small-scale composting techniques employed by micro-hauling operations also do not rely on the large energy-intensive turning or tumbling machinery employed by large composting facilities. Adding to the benefits of community-scale operations is the fact that even more energy and carbon are saved again when finished compost is used within city limits, rather than transported back into the city from a distant facility. In order to even further reduce energy expenses associated with composting, it would be sensible to carry out composting operations on even smaller scales in large cities, conceivably with each neighborhood being responsible for composting its own wastes.

Community-scale composting operations can also produce compost that is of a better quality than municipally produced compost. Municipally produced compost often contains a variety of contaminants such as plastic bags, polystyrene cups, etc. Many of these originate from curbside collections of yard rakings, which will contain debris that home owners neglect to remove. These nonorganic particles can end up in the finished compost product, which may be permitted to have as much as 2 percent of its weight comprised by contaminants. Fragments of shredded plastic bags throughout compost is visually unappealing and could result in unknown health effects (Rich, 2007). When food wastes are processed on a smaller scale by non-mechanized means, it is easier to screen out non-compostable materials. A handful of studies have demonstrated that small-batch compost is as good or superior in overall quality to industrial composts (Farrell, 2010); (Barrena, 2014).

Community composting versus AD techno-beasts

One aspect of composting that makes it fundamentally different from handling trash or recycling is the relatively decentralized and horizontally distributed nature of the composting process. Composting can be carried out on scales as small as apartment worm bins, backyard compost piles, or neighborhood-level composting with minimal start-up costs and no regulatory requirements. This stands in stark contrast to recycling facilities, landfills, and incinerators, which all entail significant capital investments, permitting, and operational expenses. There are no "micro-scale" landfills or incinerators (at least that operate legally – plenty of people illegally bury or burn trash in their backyards) with most smaller-scale municipal dumps having been shut down on account of economic pressures and strict environmental regulations (Wade, 2018). Even medium- to larger-size composting facilities (on local farms, for instance) operate at scales far less intensive than a major landfill facility and as such are simpler to permit and operate (Figure 6.6).

The decentralized nature of the organics recycling industry therefore makes it difficult for any one single entity (or handful of entities) to control and monopolize it. Certainly, there are mega-scale composting operations run by corporate waste giants, but these typically handle the only wastes generated by large-scale facilities that are delivered in bulk – it's likely not cost-effective for them to go after widely diffuse, largely residential, non-source point food wastes that would need to be collected by trucks in densely populated neighborhoods. Nimble community-scale composters on foot, bikes, and small cars are simply much more efficient at doing this.

Figure 6.6 Worm Composting

Source: Scott Kellogg

Furthermore, mega-scale composting facilities generally need to be in remote locations on account of their need for large amounts of cheap land, less stringent regulations, and low surrounding population density (fewer people to complain about noise and smells resulting from high-volume industrialized operations).

Big waste industries, however, may be attempting to consolidate the composting market by constructing large-scale anaerobic digesters. As sealed, centralized, intensive, technologically sophisticated, and grid-tied systems, anaerobic digesters permit corporations and their municipal partners to control the flow paths of organic waste resources and to profit off the gases and residues generated by them. With sewer lines and food waste trucks running into them, and with municipal gas lines running out, anaerobic digesters facilitate linear urban ecological processes and exacerbate metabolic separation. By taking the control and regenerative potential of organic wastes out of the hands and soils of local diverse communities, anaerobic digestion systems redirect and concentrate this resource to be under the management of a powerful few.

Anaerobic digestion is a composting technology that relies upon methanogenic (methane producing) bacteria to break down organic materials under anaerobic (low-oxygen) conditions, typically in a sealed container. The methane gas produced through anerobic digestion is then collected, purified, and sold, by being pumped back into either municipal gas lines or tanker trucks. Residual composted solid materials, or slurry, resulting from the AD process can then theoretically be used as fertilizer following a subsequent aerobic composting treatment that breaks down toxic compounds resulting from anaerobic decomposition (this material, however, can often become contaminated when mixed with sewage effluent).

Anaerobic composting, like aerobic composting, can be carried out on a variety of scales. It is possible to build a small functioning biogas digester using only a five-gallon bucket and water jug (I use such a device at the Radix Center to demonstrate the principles of AD and to demystify methane production). In many parts of the Global South, small- to mid-size village-scale AD systems convert animal manures into combustible gas, providing energy autonomy for their communities. AD, in certain contexts, could be considered among a tool kit of renewable energy options. I mention this to make it clear that I am not opposed to AD technology *per se* but am critical of the scale and applications of large-scale systems being built in Western cities and the threat they pose to community control of organic resources. I wish to also be clear that given a choice between AD, landfilling, or incineration, AD is a greatly preferable method for managing organic wastes. I do believe, however, that this is a false choice and that the option for decentralized and community-scaled approaches to managing organic wastes is vastly preferable still and seldom considered with seriousness. Governmental agencies seem to have an overwhelming preference for "big" solutions to wicked problems that are conducive to centralized oversight and management. Decentralized approaches, while typically supported, are not regarded as having serious potential to address large-scale waste issues.

What follows is a detailed comparison of centralized anaerobic digestion and community-based composting. In illustrating the differences, I hope to make the case for the numerous social, cultural, economic, and ecological benefits of community-based strategies.

Spatial considerations

Anaerobic digesters occupy a concentrated footprint. From the perspective of urban planners advocating high-density development, this may be viewed as a plus. Very likely, however, AD systems would be constructed in low-income areas of a city already struggling with environmental justice challenges. They return very little or nothing of value to the surrounding community in terms of either jobs or soil remediation. In this sense, AD is comparable to industrialized waste processing systems such as incinerators, transfer stations, or sewage treatment plants.

Community-based aerobic composting (CAC) systems, in contrast, are decentralized and widely distributed. CAC can be carried out in multiple, small vacant lots, which many cities have an abundance of. Many of these vacant lots, suffering from damaged, degraded, or contaminated soils, are perfect for CAC operations. The very act of composting is a form of in situ soil remediation:

- Compost piles drain nutrient-laden leachate which can biostimulate microbial populations in the soil.
- Finished compost can then be spread, creating a physical barrier between people and contaminated soils.
- A layer of compost can prevent contaminated soil from drying and becoming dust-borne.
- Compost neutralizes soil pH so as to limit the mobility of toxic metals.
- Compost reduces the biological availability of lead.

Finished compost may also be then used to fertilize community garden beds, which may occupy the same site as the composting operation. While difficult to precisely calculate, most cities (particularly shrinking cities) have enough vacant properties and unemployed population that they could conceivably be processing all their organic wastes through CAC approaches, precluding the need for the construction of AD facilities. This potential is often overlooked. While official agencies may support in theory the idea of community-based or micro-scale composting, it is rarely considered as a viable solution to large-scale waste problems. This, I believe, is a failure to appreciate the ability of CAC to be horizontally distributed. While one small vacant parcel may be able to compost a quantity of waste that is insignificant in the big picture, multiply that one parcel one thousand times and the waste processing potential increases proportionately.

Technological and economic comparison of AD to CAC

As technological systems, AD and CAC operations differ greatly. AD is contained in sealed vats with a complicated and tightly coupled series of pipes, valves, pressure gauges, scrubbers, flares, and microbes that must all function perfectly in coordination with one another for the system to function. Should any one part of the system break, it would result in a cascade of failures that would affect the entire operation. In a worst-case scenario, explosion or toxic release is possible. As such, AD requires continuous oversight, monitoring, and repair to prevent systemic collapse. The people employed in this monitoring are typically a handful of highly trained technologists who are likely not drawn from local low-income neighborhoods. Their skills and knowledge are then not typically recirculated back into the surrounding communities whose wastes feed the AD. AD systems are also very expensive to build and operate. While the cost varies considerably according to the size of a facility, total costs in the multiple millions range are typical for municipally run systems (Fries, 2018).

CAC operations, in contrast, consist of simple open-air piles, mounds, or windrows (rows) of composting material that require minimal, if any, mechanical processing, with most of the loading, mixing, and turning being done by human power. For this reason, many local jobs can be created with minimal training in everything from food scraps collection, compost management, and fertilizer sales. Loosely coupled complex living systems, compost piles produce and respond to their own generated negative feedback loops and as such are self-limiting. For instance, a properly made aerobic compost pile will seldom exceed 160 degrees Fahrenheit as the microbes responsible for generating the heat will slow as temperatures near that mark, preventing a runaway reaction (interestingly, the only compost piles to explode or catch fire are poorly maintained mega-size piles that went anaerobic and produced a build-up of explosive methane inside themselves – an extremely unlikely scenario at the CAC level). Close monitoring of compost piles is possible (it will result in finished compost faster); however, it is not required. In fact, once they are set up with the proper ratio of carbon to nitrogen, it is possible to simply walk away from a compost pile, confident that the indigenous community of microbes, worms, fungi, and so on will carry out the composting process on their own over time. In short, benign neglect toward CAC will still generate finished compost, while with AD it could be catastrophic.

A few other technically related points of difference between AD and CAC are worth mentioning. One is that anaerobic composting, unlike aerobic composting, requires temperatures in the range of 75–90 degrees Fahrenheit in order to sustain the microbial processes. In tropical climates this does not create a problem, however, in cool climates anerobic digesters function either seasonally or by being supplied with significant supplemental heating (often made by burning the gases they themselves are generating – thus negating the benefit of having them in the first place). A second issue is loading. Anaerobic digesters require a nearly constant and consistent stream of organic matter in order to function – they are intolerant of any variation in loading as too little material will starve the microbes and too much will overwhelm its capacity. AD's inflexibility in loading requirements creates an incentive to be continuously "feeding the beast". Were a municipality critically dependent on the energy produced by an AD, a situation might be created where there was a legal imperative to be directing organic waste flows to the digester (especially if they have invested millions in its construction).

Furthermore, the organic material entering an AD must be turned into a slurry – essentially liquefied – thus requiring significant mechanical pretreatment. Proposals have been put forth that would require residents to install insinkerators in their homes in order to liquefy food wastes that would then be transported via sewer to AD facilities.

CAC systems in contrast are capable of functioning year-round in cold climates. This is because they are heat generating – producing enough warmth to sustain composting processes inside themselves even during winter months. The waste heat, which is otherwise escaping to the air, may even be captured through heat exchanger systems and used to warm nearby structures – a process called bio-thermal heating. Aerobic compost piles can additionally handle considerable fluctuations in their loading requirements. Putting more or less organic matter into a compost pile will not significantly impact its function – as long as the proper balance of carbon to nitrogen is maintained, a compost pile can ramp up or slow down accordingly to its throughput.

Charting a binary

Table 6.2 Charting a Binary

Anaerobic Digesters	*Community Aerobic Composting*
Centralized/single location	Decentralized/multiple locations
Municipal/corporate control	Community control
Vertically distributed – single mega-scale	Horizontally distributed – multiple small-scale
Technologically sophisticated/complicated, tightly coupled/closed system	Technologically simple/loosely coupled/ open system
Employ a handful of highly trained specialists	Large numbers of employees with minimal training
End product slurry requires additional aerobic processing	End product compost ready for soil application
Produces combustible methane	Produces recoverable heat through microbial metabolism
No direct benefit to local soils	Benefits the health and decontamination of soils in the composting facility
Food wastes must be mechanically liquified before being placed in the digester	Food waste require minimal/no processing before being composted
Requires supplemental heating during cold periods	Provides supplemental heating during cold periods
Potential risks of malfunction, explosion, toxic releases	Minimal risks
Requires continual and consistent loading volumes – little capacity for variation in flow through	Can accept unusually large or small amounts with little impact on overall function
Microbially specific, monitored, reductive, closed, fragile	Microbially open, diverse, complex, self-organizing, resilient. Multispecies composting ecosystem

Carbon reductionism

When composting remains an aerobic process, the predominant gaseous by-product is carbon dioxide (only minuscule to zero amounts of methane are produced through proper aerobic composting). On account of growing awareness and activism related to climate change, however, an overly reactive and erroneous belief has emerged among some that carbon dioxide, regardless of its emissions source, will contribute to climate change and its production should therefore be avoided. Many are not aware of the distinction between biogenic carbon (produced from biological processes currently on the planet's surface – breathing, defecating, dying) and fossil carbon (originating from fossil fuels – ancient plant matter-turned oil/coal/gas and brought to the planet's surface). The latter plays a far greater role in increasing atmospheric carbon dioxide concentrations, while the former is part of the planet's fixed biogeochemical carbon cycle.

This is partly to blame upon a phenomenon described as "climate fundamentalism" or "carbon reductionism"(Eisenstein, 2018), defined as a tendency to reduce the analysis of climate change to quantifiable units of carbon dioxide and their greenhouse gas equivalents. Although this simplistic methodology is driven by an understandable urgency to translate the uncertainties of climate science into simplified policy-compatible metrics, it can have the unintended consequence of putting up "complexity-blinders" to the multiple and dynamic causes and impacts of climate change. Prescribed solutions of carbon reductionism tend to be technocratic, calling for source-point carbon dioxide lowering without providing a deeper critique of the systemic economic and political forces responsible for carbon dioxide pollution. For instance, carbon-reductionist tunnel vision might view nuclear energy as a viable solution to climate change, with nuclear power plants, at the point of electricity generation, having zero carbon emissions. By focusing only on carbon dioxide at the point of production, however, the enormous environmental and political hazards of nuclear energy are overlooked (as is the considerable amounts of carbon dioxide emitted in the entire nuclear fuel production cycle). Furthermore, while climate change does indeed pose an immediate and serious existential threat to humanity's survival, it is only one of many other environmental crises looming in our future (energy depletion, biodiversity loss, and environmental toxicity, among others). Solutions to climate change enacted today should not leave us worse prepared to confront future challenges tomorrow.

A possible consequence of carbon reductionism could be the regulation and prohibition of carbon emissions, regardless of whether they are of biogenic or fossil origins. As efforts are already underway to control (and possibly privatize) the planet's carbon cycle (Heynen, 2005), such a future scenario may not be entirely outside the realm of possibility. Conceivably, carbon could become either a controlled neoliberal commodity or strictly regulated under an enviro-technical authoritarian regime. In anticipation of this possibility, the rights of citizens to compost and produce biogenic carbon should be proactively safeguarded. Attempts to prohibit backyard composting on the pretense of carbon regulation might be packaged with a municipalities' desire to control their city's waste stream. Plans to illegalize backyard composting by citizens were once explained to me in a personal communication by an advocate of centralized anaerobic digester technologies. Their belief was that citizens should be required to direct all their organic wastes into a municipal biogas digester so that methane could be generated to fuel the city's police car fleet. This instance bears similarities to laws in communities in the American West, where decentralized rooftop rainwater harvesting is prohibited on the grounds that municipalities own the rights to any and all rainfall landing on their boundaries (Kwasniak, 2009). I believe it is important to defend the individual rights of citizens to engage in both small-scale rainwater harvesting and in backyard composting, as these are both essential technologies for developing urban ecological resilience and can both be thought of as resources of the commons.

Building a small-scale anaerobic composter

The reader may find it strange that after spending multiple paragraphs critiquing anaerobic digesters, I will now provide instructions on how to build one. The critical difference, like with many things, is a matter of scale. The micro-scale anerobic composter discussed here will produce no more than five gallons of combustible biogas, barely enough to warm a pot of water, let alone boil one. Due to its size, this biogas digester (Figures 6.7 and 6.8) will not divert significant volumes of organic matter away from aerobic compost piles and will serve only as an experimental device that can familiarize the public with the basics of anaerobic digestion and its potential hazards (also importantly, the small volume of gas contained in the system creates little risk of explosion or asphyxiation). Pedagogically, I also find them useful for demystifying the process of producing natural gas. Demonstrating that methane can be created in a swamp, intestinal tract, or a five-gallon bucket in the backyard takes power away from the technologically sophisticated and polluting infrastructures of hydraulic fracturing and pipelines.

What you will need:

- A five-gallon water jug (the kind used in office water dispensers)
- A five-gallon bucket (slightly larger in diameter than the jug)
- Bung cap with ¾ inch threaded insert
- ¾ inch threaded hose bib
- Tube of silicone sealant

Figure 6.7 Biogas Digester

Source: Scott Kellogg

Figure 6.8 Inside of Digester

Source: Scott Kellogg

What to do:

- Step 1: cut the bottom off of the five-gallon water jug.
- Step 2: place the bung cap over the top of the five-gallon jug and seal with silicone.
- Step 3: drill a half-inch diameter hole into the threaded insert of the bung cap.
- Step 4: thread hose bib into threaded insert.

Fill the five-gallon bucket with a compostable material slurry. The carbon to nitrogen ratio in anaerobic composting is the same as in aerobic composting – 30:1. Materials can include manure, food scraps, leaves, etc. Make sure the materials are well chopped up and mixed with water – increasing surface area will accelerate the process. Open the valve on the five-gallon jug and push it down into the bucket until it touches the bottom. Slurry will fill the jug as it descends. Close the valve and place the assembly in an environment that is between 75 degrees Fahrenheit and 90 degrees Fahrenheit (ideally outdoors).

As bacteria consume the organic matter in the digester, they will produce gas as a by-product that will cause the jug to rise. The first time it rises, it will be filled primarily with non-combustible carbon dioxide. Open the valve and press the jug all the way to the bottom to dispel the carbon dioxide. By the second time the jug has risen, methanogenic bacteria will have likely become active and producing methane gas. While exerting positive pressure (pushing downwards), open the valve and hold a lighter flame near the opening on the hose bib. The gas should now burn and be visible as a blue or clear flame.

The digester may produce another batch or two of gas but will eventually run out of fuel. At this point, remove the jug and dump the spent compost slurry into an aerobic compost pile. The spent slurry must go through an aerobic composting phase before it is safe to use as an agricultural amendment – do not dump uncomposted anaerobic slurry onto garden plants!

Radix's Community Compost Initiative

The Radix Ecological Sustainability Center runs a program called the Community Composting Initiative (CCI). The CCI is a compost collection service for people who for various reasons lack the time, space, and/or energy to compost in their backyards. Food wastes are collected from residents and businesses on a weekly basis for a small fee and are brought to the Radix Center. Wastes are mixed with wood chips (obtained for no charge from private companies) into piles where the food scraps compost passively over a period of months.

Residents are supplied with a two-and-a-half-gallon compost container with a carbon-based filter on its lid designed to scrub out odors from the food waste. The container is lined with a compostable bag made of corn starch. The function of the bag is to keep the bucket cleaner and to make it easier to dump the contents – particularly in the winter when food waste might otherwise freeze and stick to the sides of the container. The compostable bag is added to the compost pile along with the food waste where it degrades.

Once a week, CCI members put their compost containers on their front steps or porch where they are collected and dumped into eighteen-gallon plastic storage bins. Eighteen gallons is the largest size container that is used as wet food scraps are heavy, and it is difficult for one person to transport and dump a larger container.

In warm weather, the food scraps are put directly into a compost pile as soon as possible. If the food scraps sit in sealed plastic containers for too long they begin to break down anaerobically, attract flies, and create excessive odors. It is less critical to quickly process the containers in the winter. When temperatures are subfreezing the food scraps remain frozen and can be processed at a more leisurely rate. Proper management of community-scale composting operations in cities is especially important, particularly when it comes to odors and the perception that the operation is attracting pests. Foul odors and the perception of pestilence will serve only to create a bad reputation for community composting and will result in pressure to ban composting from inside city limits.

Some of the best practices for composting include:

- Only accepting plant-based materials from customers. It is certainly possible to compost putrescent or animal products like meat, dairy, and cheese, but if not properly managed they can result in serious odor and pest issues. By limiting collections to vegetables, it is possible to control the quantity of putrescent waste yet still retain the ability to occasionally compost the body of chicken or rabbit that has died. In addition to vegetables, it is possible to accept coffee grounds, tea bags, paper towels and napkins, shredded newspapers, flowers and houseplants, hair or fur, holiday wreaths, jack-o-lanterns, and eggshells.
- Timely processing of the compost and ample additions of carbon-based materials are necessary to suppress odors. It is always preferable to add too much carbon and slow the composting process versus having too many nitrogen-based materials that will produce odors.
- Wrapping compost piles in hardware cloth or chicken wire will inhibit access to the pile by mice or rats. If vermin are a serious issue, it may be necessary to compost in sealed containers, although this is less desirable as it reduces the oxygen levels in the compost pile.
- Finally, layering straw or hay around the outside of a compost pile, making it resemble a cylindrical hay bale, will create a visually appealing screen, preventing anyone from seeing food waste. An additional benefit is that the straw layer helps to insulate the pile and thereby accelerates the composting process.

It is a commonly held misconception that compost piles need to be regularly turned. While turning a pile will accelerate the composting process, it is not necessary. Turning a compost pile mixes its contents, moving organic material from

Figure 6.9 Community Composting Bike Hauling

Source: Scott Kellogg

the cooler edges of the pile into its hotter, more microbially active core. The most intensive turning methods can produce finished compost in as little as three weeks. While it may produce finished compost more quickly, a turned pile also requires much greater labor inputs (Figure 6.9).

Unless a composting operation is being run where space is incredibly limited or a very high volume of material is being processed, it may be preferable to use the static or passive composting method. With this technique, no turning is required. Food wastes are layered with wood chips and allowed to break down over time through the actions of microbes, insects, and worms. Depending on the season, the passive process may take 8–10 months instead of a few weeks. (Compost breaks down faster in warmer weather.) In this regard, passive composting is time and space intensive while actively turned piles are labor-intensive.

Compost and soil remediation

An appropriate use for contaminated or brownfield sites is to repurpose them as small-scale composting facilities. The reason for this is that the addition of compost to contaminated soil, as well as the act of composting itself, can assist in the reduction or immobilization of soil-based pollutants (Ceccanti, 2006); (Farrell, 2010). The benefits of composting on contaminated sites include:

- Compost acts to encapsulate contaminated soil, creating a physical barrier between people and the soil and prevents contaminants from becoming dust-borne.

- It creates a microbially enhanced environment that can facilitate the degradation of organic pollutants.
- It neutralizes soil pH so as to reduce the solubility of both cationic and anionic metals.
- Lipophilic metals such as lead can bind with the compost, thereby reducing their biological availability.

After a sufficient amount of finished compost has been created on a site (enough to create six to twelve inches of soil), it can be spread about and used for agriculture purposes.

Chickens as nature–culture boundary objects and chicken-assisted composting

Philosopher Bruno Latour frequently struggles with the dualism of nature and society, lamenting the absence of any single word that is sufficient for "reunifying" the two. Latour uses the idea of nature–culture as a hypothetical space where the dichotomy of subject–object, nature–artifice can at least be temporarily resolved in an attempt to move us beyond the constraints of "modernity". I would put forth the idea of the chicken as a symbol of that imagined space, representing both the potential positive and negative aspects of nature–culture. Going further, I believe that people's attitudes about chickens are in many ways reflective and indicative of their ideas and feelings related to modernity, nature, and society and for this reason make them a fascinating subject of study.

In my time as a sustainability activist, I have been shocked by the extreme emotions that are brought out in people over the idea of chickens in the city. I've seen ugly political battles take place over the issue, with people arguing passionately both for and against them. I've often wondered what is it about chickens that bring out such charged feelings in people? They are, after all, considerably less potentially dangerous than just about any dog and create less nuisance than many feral cats, both of whose belonging in cities is largely unquestioned. Why then are chickens so contentious? My personal belief is that people object more to what chickens represent, rather than to chickens themselves.

Chickens have a long history of domestication, dating back several thousand years to Southeast Asia. From this time until the present day, there has a been a productive symbiosis and coevolution of humans and chickens, with them providing a hugely important source of meat and eggs, while their populations in turn today number in the multiple billions. Benefits of keeping chickens extend beyond their value as food, however. They are effective at managing pests (chickens are omnivorous – in addition to grains their diets include insects, lizards, snakes, and mice), tilling and weeding soils, and converting food wastes into garden nutrients. Despite there being a long history of people keeping chickens in cities, they have largely been absent from the urban ecosystem for most of the past century. Their exile was in many cases due to laws passed that targeted recent immigrants, a population more likely to be relying on chickens for sustenance. Today, chickens are commonly associated with rural living, which can evoke both positive and negative responses from people (the term "yardbird" is a southern expression for chickens, reflecting the fact that chickens commonly range freely about people's homes). I have encountered several older Albany residents who have remarked that the only place in a city where a chicken belongs is in a grocery store – they perceived cities as being "clean" environments where "dirty" chickens did not belong. These attitudes may very well have come from people having once been responsible for taking care of chickens, which if not managed correctly have the potential to be odorous, just as any other animal would be.

Despite negative perceptions of chickens, several American cities have passed laws in recent years that permit residents to keep chickens. This change has been driven by citizens aware of the multiple benefits that chickens can provide to urban communities. In addition to the obvious benefits of providing eggs and helping to improve a neighborhood's food security, chickens also can play a significant role in organic waste reduction.

Chickens are little walking garbage disposal machines. Omnivorous by nature and undeterred by bits of rot or mold, they are happy to digest a whole range of food wastes deemed to be unfit for human consumption. They are effective at rapidly reducing the volume of food scraps produced by either a household or a community-scale composting operation and upcycling the nutrients into both eggs and chicken manure, a fantastic fertilizer. Radix maintains a flock of approximately twenty chickens and twenty ducks (collectively referred to as microlivestock) that assist greatly in the community composting initiative, consuming around 90 percent of the food waste brought in is fed to the birds (the remaining 10 percent goes into static composting piles). Carbon-based material (wood chips and hay) is frequently spread about the chicken run to cover manure and to ensure a 30:1 carbon to nitrogen ratio. Once the material accumulates sufficiently, it is dug out, piled up, and permitted to age for a period separately (around three to four months). After it has sufficiently aged, the thoroughly broken down compost is then spread over garden beds where it makes a premium soil nutrient (Figure 6.10).

The benefits of chicken-assisted composting can be summarized as such:

- Rapid reduction of food waste volumes
- Generation of eggs that can be sold or consumed
- Production of manures to be used as a garden amendment
- Elimination of the need to purchase commercial feeds
- Chicken education (fun, interactive way to teach about cyclical food/waste systems)

Conclusion

Despite the critical role that compost and composting plays in building regenerative and just waste, soil, climate, and food systems, it remains a surprisingly under-studied and under-theorized realm of research. The dominant discourse surrounding organics recycling is led by technical and business experts, often looking

Figure 6.10 Chicken-assisted Composting

Source: Scott Kellogg

to capitalize off a new frontier of waste that is coming under greater public scrutiny. It is critical that urban ecological activists insert themselves into conversations about composting and insist that the needs of environmental justice communities are central to compost policy and planning. With the potential to mitigate climate change, provide jobs, regenerate toxic soils, and bolster food security, compost is too valuable of a resource to hand away to mega-scale waste handling corporations who stand to profit off compost's control and consolidation. Composting must remain a process that is decentralized, horizontally distributed, and citizen controlled. Ensuring this will allow communities to close the loop on linear urban metabolic processes and begin to cultivate urban ecosystem justice in their neighborhoods. It is my wish that the field of critical compost studies matures and that practices and methods for compost justice are further refined and broadly shared.

Note

1 By focusing on composting, I wish to be clear that in no way do I intend to dismiss or ignore the long and important history of environmental justice struggles related to waste in urban areas. Battles over the placement of waste infrastructure, as well as a city's responsibility to collect waste, have been and continue to be fought in inner-city neighborhoods throughout the world. Low-income communities have had to bear the brunt of exposure to noxious and toxic emissions from landfills, incinerators, sewage treatment facilities, and transfer stations: waste facilities to which the wealthy contribute to disproportionately. This has only intensified in the Global South as exported plastics and electronic waste disposal disproportionately impact impoverished communities. While these ongoing struggles are of critical importance to establishing social and ecological justice in cities, they are largely (and by necessity) reactive in nature. In keeping with the primary focus of urban ecosystem justice to focus on the equitable distribution of both environmental harms and amenities, and to explore possibilities for proactive environmental justice action with a strong citizen-centered component, I have chosen to focus on the topic of organics composting.

Works cited

Agyeman, J. (2005). *Sustainable communities and the challenge of environmental justice*. New York: New York University Press.

Anastas, P. T. (2000). The role of catalysis in the design, development, and implementation of green chemistry. *Catalysis Today*, 55(1), 11–22.

Attanayake, C. P. (2015). Potential bioavailability of lead, arsenic, and polycyclic aromatic hydrocarbons in compost-amended urban soils. *Journal of Environmental Quality*, 44(3), 930–944.

Balée, W. L. (2006). *Time and complexity in historical ecology: Studies in the neotropical lowlands*. New York: Columbia University Press.

Barrena, R. e. (2014). Home composting versus industrial composting: Influence of composting system on compost quality with focus on compost stability. *Waste Management, 34*(7), 1109–1116.

Campbell, S. (1998). *Let it rot! The gardener's guide to composting*. North Adams: Storey Publishing.

Ceccanti, B. e. (2006). Soil bioremediation: Combination of earthworms and compost for the ecological remediation of a hydrocarbon polluted soil. *Water, Air, & Soil Pollution, 177*(1), 383–397.

Eisenstein, C. (2018). *Climate: A new story*. Berkeley, CA: North Atlantic Books.

EPA, U. (2017). LMOP – Landfill Methane Outreach Program. *United States Environmental Protection Agency*. Retrieved July 2, 2017, from www.epa.gov/lmop/basic-information-about-landfill-gas

Farrell, M. a. (2010). Use of composts in the remediation of heavy metal contaminated soil. *Journal of Hazardous Materials*, *175*(1), 575–582.

Fries, A. (2018, March 28). *Albany, Saratoga counties move ahead with $45M organic waste facility*. Retrieved from Times Union, www.timesunion.com/news/article/Albany-Saratoga-counties-move-ahead-with-45M-12787834.php

Gouin, T. e. (2004). Evidence for the "grasshopper" effect and fractionation during long-range atmospheric transport of organic contaminants. *Environmental Pollution*, *128*(1–2), 139–148.

Gunders, D. (2012). *Wasted: How America is losing up to 40 percent of its food from farm to fork to landfill*. New York: Natural Resources Defense Council.

Gustavsson, J. e. (2015). Global food losses and food waste. *Food and Agriculture Organization of the United Nations website*. Retrieved April 30, 2017, from www.fao.org/food-loss-and-food-waste/en/

Haraway, D. J. (2016). *Staying with the trouble: Making kin in the Chthulucene*. Durham, NC: Duke University Press.

Herrmann, D. L., Schwarz, K., Shuster, W. D., Berland, A., Chaffin, B. C., Garmestani, A. S., & Hopton, M. E. (2016). Ecology for the shrinking city. *BioScience*, 66(11), 965–973.

Heynen, N. a. (2005). The neoliberalization of nature: Governance, privatization, enclosure and valuation. *Capitalism Nature Socialism*, *16*(1), 5–8.

Hoornweg, D. P.-T. (2013). Environment: Waste production must peak this century. *Nature News*, 615.

King, F. H. (2011). *Farmers of forty centuries or permanent agriculture in China, Korea and Japan*. Folkestone: Global Oriental.

Kwasniak, A. J. (2009). Right to rainwater: A cloudy issue. *Windsor Review of Legal & Social Issues*, *26*, 105.

Lefebvre, H. (1996). The right to the city. *Writings on cities*, 63181. Malden, MA: Blackwell Publishers.

McClintock, N. (2012). Why farm the city? Theorizing urban agriculture through a lens of metabolic rift. *Cambridge Journal of Regions, Economy and Society*, 191–207.

Montgomery, D. R. (2012). *Dirt: The erosion of civilizations*. Berkeley, CA: University of California Press.

Papargyropoulou, E. e. (2014). The food waste hierarchy as a framework for the management of food surplus and food waste. *Journal of Cleaner Production*, *76*, 106–115.

Protection, M. D. (2019). *Per- and Polyfluoroalkyl Substances (PFAS)*. Retrieved from Maine Department of Environmental Protection, www.maine.gov/dep/spills/topics/pfas/index.html

Rich, D. (2007). *Taking out the trash: Municipal green waste needs cleaning up*. Special to The Chronicle. Retrieved from www.sfgate.com/green/article/Taking-out-the-trash-Municipal-green-waste-needs-3301750.php

Schaffer, G. D. (2016). *Building utopia with tomorrow {u2019} s trash: Formal and informal infrastructures for organics recycling in New York City*. Dissertation, Rensselaer Polytechnic Institute.

Sierra. (2010). *Landfill-gas-to-energy, Sierra Club Report on*. Prepared by the Sierra Club LFGTE Task Force. Retrieved September 2, 2017, from www.sierraclub.org/sites/www.sierraclub.org/files/landfill-gas-report.pdf

Swyngedouw, E. (2006). Metabolic urbanization: The making of cyborg cities. In *In the nature of cities* (pp. 36–55). London: Routledge.

Wade, L. (2018, December 15). Local town dumps a thing of the past due to high cost of government regulations. *Des Moines Register*. Retrieved from www.desmoinesregister.com/story/news/2018/12/15/local-town-dumps-thing-past-due-high-cost-government-regulations/2322022002/

Weber, C. L. (2008). Food-miles and the relative climate impacts of food choices in the United States. *Environmental Science and Technology*, *42*(10), 3508–3513.

7 Urban biodiversity justice

According to modernist philosophy, cities are "human only" spaces built by and for the exclusive use of homo sapiens – clean, sterile artifacts of human imagination that symbolize humanity's separation from nature. Aside from cultivated garden plants and a handful of companion or work animals whose presence is tolerated, non-human life in cities is regarded as matter out of place: inconvenient and unwanted at best and at worst, reviled and exterminated (Houston, 2018).

A new awareness is emerging, however, which problematizes the overly simplified nature–society dualism discourse. It brings to light the fact that cities have always been assemblages of multiple species, the product of numerous human and non-human coevolutionary processes (Alberti, 2008). Critiques from disparate literatures, including zooarchaeology, animal studies, urban environmental history, multispecies ethnography, and the post-humanities have arisen to further unsettle the notion of urban human exceptionalism that has become ingrained in popular consciousness.

Running parallel with the multispecies analysis, the subject of urban biodiversity has matured greatly within a broader framework of the urban ecological sciences. It has grown from an "ecology in the city" paradigm study of "remnant populations" to a field that looks at the numerous ecosystem services provided by non-human life in the city and how human/non-human cohabitation can be enhanced through appropriate design. There is an emerging understanding that cities may often support greater levels of biodiversity than surrounding areas (particularly industrial agricultural zones) and can function as refuges for certain threatened species. As many of the world's fastest growing cities are on the edges of some of the world's biodiversity hotspots, research is being conducted on how urban development and biodiversity conservation can be comanaged, supported empirically by research fields such as "reconciliation ecology". Urban biodiversity has now become an area of global urban policy, planning, and governance, documented in the extensive "Cities and Biodiversity Outlook" report drafted by the Convention on Biological Diversity (Elmqvist, 2013).

Mainstream studies on urban biodiversity, however, typically examine the issue through a highly reductionist lens: focusing on quantifiable data that can be extrapolated into generalizable best practices, applicable to planning and architectural strategies in any city in the world.[1] Seldom is attention paid to questions of

DOI: 10.4324/9780367858711-8

politics and power as they apply to urban biodiversity. Which populations benefit from the ecosystem services provided by non-human life in cities, and are those populations represented in decision-making processes related to biodiversity? How might certain people's survival be closely linked with the well-being of urban non-humans? How can justice, equity, and fairness for human urban residents be furthered by forming mutually symbiotic interspecies partnerships? Going further, it can be asked whether non-humans have just as much of a "right to the city" as humans do – are squirrels, birds, insects, amphibians, fish, fungi, weeds, and microbes as equally entitled to life and liberty in the urban ecosystem as humans?

Admittedly, answers to these questions are hard to come by on account of the paucity of research related to them. For a long time, concerns related to social justice and biodiversity conservation have been at odds with each other, with considerable tensions between communities advocating for either. There are many examples of conservationists, arising from "Edenic" traditions and ideologies, seeking to create wilderness preservations that are off-limits to humans, indigenous or otherwise, even when humans may historically have played pivotal roles in the creation and functioning of those ecosystems (Heise, 2016). The separation between conservationists and social justice activists is further complicated by the fact that the divide often corresponds to race and class distinctions, with more typically white, Northern, and affluent conservationists accused by the latter of caring more about animals/plants/ecosystems than they do about the suffering of members of their own species (the same critique has been similarly applied to advocates of urban greening versus environmental justice). Surely, in low-income urban communities, it can be difficult to convince someone struggling to put food on their table that they should care about the well-being of the birds and the bees living around them. This alienation is compounded in cities by the so-called luxury effect: a phenomenon where wealthier neighborhoods are observed to have higher rates of biodiversity than poorer ones (Leong, 2018).[2]

What work has been done to try and bridge the divide between social justice and biodiversity conservation? Are there concepts that can be used to reconcile these seemingly disparate movements?

One potentially bridging framework is "biocultural diversity". The United Nations Environment Programme coined the term "biocultural diversity" in 1999 to describe the "inextricable link" between the interrelated and coevolved manifestations of biological, cultural, and linguistic diversity manifesting in complex socio-ecological systems (Buizer, 2016). Primarily focusing on indigenous and traditional peoples, the study of biocultural diversity contrasts the retention of biocultural knowledge among indigenous groups against its loss as a consequence of socioeconomic modernization. With a primary focus on rural communities, urban biocultural diversity studies have been few. A few recent publications, however, have started this conversation. Focusing on changing and evolving agrobiodiversity practices in peri-urban spaces, they have put forth the key concept of "biocultural creativity" or the creation rather than the preservation of biocultural diversity (Buizer, 2016). This concept can be extremely useful for framing the rebuilding of symbiotic mutualisms between humans and the non-human life in novel urban ecosystems and for making linkages between biodiversity and social justice. Another interesting boundary concept is "multispecies environmental justice" (Haraway, 2018), which asks what environmental justice looks like when extended beyond the humans.

Discussing urban biocultural multispecies diversity justice is a complex task – there are many angles from which to explore the relationship between humans and non-humans in cities and how they relate to broader questions of equity, fairness, and justice. Rather than producing a generalizing "unifying theory" of urban biodiversity justice that strips away the important differences, nuances, and fractal micro-stories that exist in multiple locales, I'll instead attempt to explore broad patterns of these relationships and speculate how they may later lead to the development of a more robust theory.

My primary focus on urban human–non-human relationships is within what I refer to as the "zone of reciprocity". This refers to a space of human–non-human relationality located in between the highly domesticated (dogs, cats, other pets, farm animals, garden vegetables, etc.) and the undomesticated urban "wild" (non-food producing trees, wild birds, most insects, urban deer, foxes, and coyotes, for example).[3] Somewhere in the middle of this messy continuum is the reciprocal zone: a collection of species across the taxonomic spectrum with whom we engage in mutualistic symbiosis in a semi-controlled, semi-chaotic manner – a give-and-take based on an ethos of respect and interdependency that is resistant to both domination and diffusion.

My hope is that by exploring and better understanding the zone of reciprocity, urban residents can learn to appreciate how their own health and well-being (as well as that of their families and communities) are closely linked with that of non-humans. The direct, tangible exchanges with these handful of reciprocal species make it clear how benefits to humans and non-humans are shared. While the survival of humanity is of course entirely dependent upon the continued health of the global ecosphere, urban ecological alienation often causes people to be under-aware of this truth. Synanthropic relationships can provide a first step toward cultivating connection, empathy, and the awareness of reciprocity with species on the outside (wilder side) of the zone, ultimately leading to the cultivation of a broader biocultural justice and ecological literacy.

There are a number of species that while indeed synanthropic, I am not including in the zone of reciprocity. For a relationship to be considered reciprocal, both entities must benefit in some way from their exchanges. This excludes so-called nuisance species: mice, rats, rabbits, raccoons, squirrels, skunks, and opossums. Despite, or perhaps because of their success at cohabitation with humans, these animals are commonly maligned, reviled, and systematically relocated or exterminated. While I fully support changing our relationship with these species and learning ways to coexist with them that do not involve mass killing, disturbance, or poisoning, I am currently not including them within the zone of reciprocity as, honestly, I am presently hard-pressed to see how our relationships with them could become genuinely mutually beneficial and not just less toxic. This is not to say it is impossible or that models do not already exist that I am unaware of. Crows and pigeons, two other common urban "nuisance" species, have been studied for their relatively unknown and overlooked symbiosis potential (Klein, 2007). It is my hope that these two examples could serve as inspiration for creating improved human–non-human relationships with other "nuisance" species.

I'm also leaving aquatic species out of this discussion, focusing instead on terrestrial animals and plants. While it is completely possible to have reciprocal relationships with aquatic life (oysters and carp are good examples), I have discussed this at length already in the chapter on water (Chapter 4) and for the sake of simplicity will exclude them here.

So then what species are within the reciprocal zone? In addition to the aforementioned pigeons, I include silkworms (and mulberry trees), bees (both honey bees and "native" bees), black soldier flies, maple trees, wine cap mushrooms. . . . This is admittedly a very short and limited list. There are countless other species of semi-wild edible and medicinal plants growing in cities that people have reciprocal relationships with, selectively harvesting them for food and medicine while providing protection, pruning, care, fertilization in return. Numerous studies on urban foraging detail these relationships in depth ((Poe, 2014); (Shackleton, 2017); (McLain, 2014)), I will not replicate them here. Instead, I will discuss this handful of species as they are the few that I have had firsthand experience with and believe to show promise to be further developed.

Pigeons

Humans and pigeons have a fascinating and highly complicated coevolutionary relationship. They are one of the few species that on its own accord has chosen to live alongside us in the built environment and insomuch has arguably domesticated itself. Prized by people throughout history for meat, manure, and message carrying, these beautiful birds are now commonly maligned by many urbanites, unlovingly referred to as "rats with wings", and subject to massive and costly campaigns of exclusion and extermination. Despite their undeserved poor reputation, pigeons make a fascinating subject for studying human–non-human relationships in cities. Representing true "boundary objects" (Akkerman, 2011), pigeons straddle the edges of domestic and feral, resist attempts at control, and defy modernist mythologies of cities as "human only" places.

History

The pigeon or rock dove (*Columba livia*) is a bird that has lived in proximity to humans for millennia and is likely the first bird to be domesticated by humans. Native to the Middle East, parts of Europe, and east to India, the aptly named rock dove would build its nests in holes in rocky cliffs, flying short distances to find food to fulfill their diet of grains, berries, and seeds. In approximately 3000 BC, the first cities were built by people in the region. Towers made of mud, stone, and clay closely resembled the native cliff habitat of pigeons and were soon colonized by the birds. Feeding off our wastes and enjoying the reduced number of predators (raptors) found there, pigeons quickly adapted to urban living, prospering among us and often preferring the built environment to their own.

The human–pigeon mutualism further developed when people became aware of their value for both meat and manure. In order to enhance their habitat, niches were carved into towers (literal "pigeon holes") and clay pots were hung. Both served as nesting spots for the birds, a very early form of reconciliation ecology that allowed urban pigeon populations to increase. The designed habitat made it easier to catch the birds ("squab", the meat of a young pigeon, is considered a delicacy in many places to this day) and to collect their droppings in concentrated piles ("guano", the manure of pigeons, is a potent nitrogen-rich agricultural fertilizer).[4] Shortly afterwards, "dovecotes" – structures designed purely for the use of pigeons for roosting and breeding – were constructed. Dovecotes were built in a stunning variety of architectural forms, at times resembling beehives or miniature castles, which were common features not only in early cities but on European estates as well.

The ability to keep pigeons and maintain dovecotes has not always been equitably permitted. In seventeenth-century France, only those in landowning professional classes could maintain dovecotes, and the number of pigeon nests within was determined by the size of their lands. The pigeons of the nobility were despised by local peasants as they fed on and caused significant damage to their crops. The unjust nature of pigeon keeping and the resentment it brewed are cited as contributing factors to the French Revolution, after which many dovecotes were destroyed (Cooke, 1920).

Pigeons have a highly developed homing instinct that allows them to navigate back to their home roosts, even after being transported away for many miles. The exact mechanism that they use to navigate is still not completely understood by scientists, but it's believed to be a combination of magnetic sensors in their heads along with their use of visual clues. Pigeon's homing abilities were recognized very early after their domestication and were utilized for sending messages at a speed much greater than over land travel. Early pigeon message delivery systems were created where written notes could be relayed by pigeons rapidly over great distances.

Fast forward to the present day, pigeon populations are enormous in most cities, with an estimated 1–7 million birds or more living in New York City alone. The birds, largely brought over to North America by Europeans in the 1600s as a food source, have readily adapted to life in the modern metropolis (Nigro, 2012). Regarded with disgust by many, pigeons are often regarded as dirty, foul creatures that make messes and spread disease. Much money is spent trying to exterminate or control their populations, with nearly one billion dollars alone in the United States each year. The bad reputation of pigeons is largely undeserved, however. While it's true they may leave droppings, about twenty-five pounds per bird each year, (Mooallem, 2006) on buildings and statues causing some damage due to their acidity, people's concerns regarding them are mostly cosmetic. Pigeons, like all animals, may carry certain human pathogens, however, they rarely pose any significant risk to healthy individuals or to the public health at large despite sensationalistic reporting that has claimed the contrary (Haag-Wackernagel, 2004); (Jerolmack, 2008).

Today, pigeon raising is still common in cities, though not as prevalent as it once was. In many cities, pigeon coops on the roofs of row houses are a common sight. The pigeons kept in these coops are typically not the average feral pigeon but are fully domesticated breeds that have been raised specially for purposes ranging from racing to homing to pure fanciful showiness. Pigeon coops mostly are wooden structures with wire siding that protect the birds and create nesting spots for them. The owner will typically release the birds each morning, allowing them to fly about freely. Because pigeons have such a strong homing instinct, they will typically return to their coop each night, especially if nests are present and food is provided. Rooftop pigeon flying, an urban pastime that transcends racial and class boundaries, is increasingly threatened by the forces of gentrification which view it as being "messy" and unfavorable to upscale development (Berger, 2013).

Pigeons, chickens

Pigeons and chickens have been domesticated for longer than any other bird species. Pigeons, the first to be domesticated, outnumbered chickens for hundreds of years. Today, however, the inverse is true – the number of domestic chickens has greatly exceeded that of pigeons. While there are currently an estimated 250 million pigeons on the earth, chickens come in at a stunning 22.7 billion, most of which are bred in confined industrial farms for human consumption (Bennett, 2018). How can that shift, and discrepancy, be explained? Pigeons, simply stated, do not grow as big as chickens and produce fewer eggs (a hen in her prime can lay one egg a day while a pigeon may lay one or two over several weeks). Also, because they're not capable of sustained flight, chickens are easier to contain. Taken together, these factors have made chickens more amenable to concentrated, intensive farming practices and has led to their subsequent human-assisted global population explosion.

In consideration of this, why would anyone choose to raise pigeons over chickens? If their desired goal is simply to produce food, it wouldn't make much sense. Raising pigeons, however, provides benefits that make it a complimentary strategy to keeping pigeons.

Pigeons are unique among domestic animals in that they still can feed themselves through foraging, an ability that has otherwise been largely deselected in other domestic species. In a dovecote system, pigeons are provided shelter and nesting sites (and wintertime food) but are still allowed to fly freely in search of food, returning to the nest daily to deposit manure and rear their young. In urban applications then, when permitted to forage, pigeons could be thought of as "free-range" urban birds who uses the entire city as their "pasture".

Chickens, by comparison, are far more dependent on having food provided to them by humans. In an urban area, there are few areas where it's possible to raise free range chickens. Most chickens will need to be put in an enclosure, in order to protect them from predators and to keep them away from gardens, which they'll destroy if allowed to have access to them. In an enclosed area, they will very quickly strip it of vegetation. Aside from the occasional bug that hops through, enclosed chickens will be entirely dependent on food being brought to them.

By raising chickens and pigeons together, it is possible to engage in both intensive (chicken) and extensive (pigeon) parallel and complimentary agricultural strategies. Pigeon and chickens can even share the same coop – a secure structure that can protect them from nocturnal predators.[5] Pigeons will however need separate nesting boxes high up in the coop that chickens cannot access – chickens would likely disturb a nesting pigeon and would probably eat their babies too!

A concern that immediately arises when discussing free range pigeons is whether they are accumulating toxins in their bodies by foraging in potentially contaminated areas. One study, examining the potential use of pigeons as bioindicators of lead toxicity, has documented how pigeon blood lead levels correlate with those of children in respective New York City neighborhoods (Cai, 2016). This would suggest that there may be some concern about eating free range urban pigeons – blood lead levels are not necessarily indicative of levels in muscle tissue. As use of pigeon manure as a fertilizer is more common than eating pigeon flesh though, I have personally run an analysis of pigeon manure with an X-ray fluorescence machine and had detected negligible amounts in the sample (1.2 parts per million). More research in this area is warranted.

Pigeons at Radix

The Radix Center provides a home to a sizable flock of pigeons which share a coop shelter with our chickens. The pigeons use the same door entrance as the chickens, but have designated nesting boxes built at the top of the coop where they are protected from the chickens (adult pigeons and chickens get along just fine, however chickens will attempt to nest in pigeon boxes and will displace pigeons and eat their eggs in the process – the pigeon boxes are placed at a height unreachable by flightless chickens and are additionally secured with entrance reducers that exclude chickens). Pigeons build their nests in these boxes, gluing sticks and twigs together with their own manure, where they raise their young.

Many people are surprised to learn that the pigeons were initially intentionally placed at Radix and are not just feral street pigeons that moved in on their own – pigeons on purpose, as we say. Our pigeon colony began with a group of ten birds purchased from Broadway pigeon supply in Brooklyn at six dollars each. They were an odd mix of colors and breeds, some of them having "Jesus saves" printed on their plastic leg bands. They were kept inside the coop for two weeks before being allowed to fly free – a process called "re-homing" through which they'll accept the new coop as home base. Once adjusted, the birds were let out and were free to come and go as they choose. They'll often spend the day outside, perched on the roofs of neighboring buildings and coming back to Radix to sleep and feed. I do not provide any special food for the pigeons – they peck at the leftovers of whatever the chickens have been fed and drink from the same waterers. This food, and in addition to whatever they forage, is apparently enough for them to survive and breed – at any given time there are likely babies being raised. Over the years, the population of pigeons has fluctuated considerably. Predator pressure has pushed it down at times, and at others there seem to be twenty or more birds living in the coop. I must admit that I do not keep track of individual birds, so I truly don't know if any of the pigeons I have now are descended from the original group of pigeons I started with several years ago or if they have been entirely displaced and replaced by feral pigeons. While I have never eaten any of our pigeons, I do occasionally scrape out their manure and use it as a garden fertilizer (Figure 7.1).

While their manure is valuable, my main purpose for keeping pigeons is for their educational and symbolic value. On one level, they are one of the few agents at Radix who regularly leave the boundaries of the property, interact and mix with the outside community, and return on a regular basis (honeybees may be the only other example). This helps to break down (at least my own) perceived feelings about the "permeability" of the space – energetic and material exchanges with the world outside Radix's fence line promote mixing and make it feel like less of an "isolated" system. On an educational level, I routinely enjoy bringing a pigeon out for people to pet or hold. Despite regularly being around them, very few people have ever had any direct contact with a pigeon. By allowing people to interact with one, they may question their own internalized attitudes about these "rats with wings" and hopefully better appreciate these beautiful and underappreciated birds. Pigeons also travel well. When visiting local elementary schools, I have brought pigeons with me as a way to teach about urban biodiversity. After giving students an opportunity to pet the bird, we then collectively release it and watch it fly (typically) back in the direction of Radix. When the same students later come to visit, they can once again see the same bird. In this way, the pigeon provides a way for children to think differently about the spatial relationship between Radix and their school.

Figure 7.1 Inside of Dovecote

Source: Scott Kellogg

Silkworms/mulberries

Silkworms (*Bombyx mori*) are an insect that has been raised for many centuries for their use in producing silk fibers. While more domestic than semi-wild (arguably silkworms and honeybees are the only two domesticated insect species – a result of the extent of genetic alteration they have undergone through selective breeding), I include them within the zone of reciprocity because of the obligate symbiosis between silkworms and mulberry trees (genus *Morus*), upon whose leaves they exclusively feed. While domestic silkworms are not naturally found anywhere in North America and depend completely on humans for their survival, mulberries are a common ruderal (disturbance tolerant) tree that grows and spreads prolifically in urban environments. The relationship between this domestic insect, wild tree, and humans makes an interesting subject for the exploration of reciprocal urban biocultural diversity.

"Sericulture", or raising silkworms for silk production, is a practice originating in China that dates back thousands of years (Barber, 1991). In it, larval silkworms are raised in protected environments and fed the leaves of mulberry trees.[6] As a larval caterpillar, the silkworm undergoes a rapid transformation from a tiny egg to a nearly three-inch caterpillar in a matter of weeks, gaining nearly 10,000 times its original weight. Once its growth is complete, the silkworm will spin a cocoon around itself and undergo metamorphoses into an adult silk moth. In silk culture, this process is interrupted by placing the cocoons in an oven or in boiling water. This kills the insect and allows people to begin unwinding the single silk thread that makes up the cocoon (about 1,000 yards in length). It is from these silk threads that silken fabrics are woven (unwinding the cocoons is a terribly labor-intensive process – one of the reasons why commercial silk production was not widely practiced in North America).[7] Were it allowed to complete its metamorphosis, an adult silk moth would emerge from the cocoon as a flightless white moth that lives only for a short period, long enough only to mate and (for females) lay eggs. Because it can't fly and contrasts so much against green leaves, it would have little chance of surviving predation in the wild and therefore poses no threat of being invasive.

While silkworms are not raised for their silk fibers at Radix, they are valued as a feedstock for animals (Figure 7.2). The mature silkworm caterpillars (or their pupae) make an excellent, high protein, fat, and calcium feed for chickens or fish (silkworm pupae are also commonly eaten by humans). By avoiding having to unwind silk cocoons, the labor costs of raising silkworms is hugely reduced. Furthermore, the droppings or "frass" produced by silkworms make a valuable fertilizer.

Apart from any utilitarian purpose, silkworms make a fantastic educational tool for children. For one, they are an extremely docile, slow moving insect that will neither bite, sting, attempt to run away, or escape confinement. Their skin is soft and fleshy and can easily be held in a child's hand. These factors alone are great for helping children to extend empathy to insects, a group of animals that have a high "yuck factor" – connection with them reduces disgust sensitivity and fosters

Figure 7.2 Silkworms

Source: Scott Kellogg

compassion toward non-humans. Beyond this, their rapid growth rate and ease of care make them an effective learning tool for children with limited attention spans.

Mulberry symbioses

Mulberries are a type of tree in the genus *Morus* consisting of several species found throughout the world. *Morus rubra*, or the red mulberry, is native to the eastern United States and is often hybridized with the non-native white mulberry. Along with the black mulberry, these three species are commonly found in disturbed locations throughout urban environments. Producing delicious berries, mulberry fruits are enjoyed by humans and birds alike and are widespread on account of the latter's dispersion of seeds. Some cities are home to large concentrations of mulberries (or "moriculture" farms) that are the surviving legacy of bygone attempts at establishing an American silk industry.[8]

I have found mulberries to be numerous enough in Albany that it is possible to collect enough leaves on my way to the Radix Center to sufficiently feed a colony of more than one hundred silkworms without having any noticeable impact on the tree's density of foliage. Regardless, I have planted them in numbers at the Radix Center for their multiple other uses. Extremely fast growing, I have placed mulberries around the periphery of the chicken run to provide shade and food for the chickens (by shaking the branches, I can cause multiple berries to drop inside the chicken's fence, which they quickly devour). Berries not eaten by chickens become food for humans (great for jams) or for wild birds (who are far better at accessing berries in the tall, thin branches). Furthermore, the proximity of the mulberry leaves makes feeding silkworms a more convenient task.

Raising silkworms

What you will need:

- One plastic bin approximately $1' \times 1.5'$ or larger, or a cardboard box of similar size
- 1000 silkworm eggs
- Source of mulberry leaves

To begin raising silkworms, start by ordering some eggs. There are several online providers who sell them, mostly for use in science classrooms (a Petri dish containing a thousand eggs should sell for around fifteen dollars). Before ordering them, make sure that mulberry trees are in bloom or that they will be shortly. When the eggs arrive, watch them closely. At first, they will just be black dots, resembling seeds. When they are ready to hatch, they will begin to turn white. When the silkworms first emerge, they will be tiny, like little black threads.

Immediately after hatching, the baby silkworm's jaws are too weak to eat mature leaves. Pick the newest, soft, and tender leaves from the tips of the mulberry's branches. These will usually be a lighter green color and can be a food for the larvae in their first couple days. Right away, you'll see the newly hatched larvae covering the leaves, slowly munching away at them. Make sure there are no water droplets on the leaves, as a larva can drown in them. The larvae get all their moisture from the leaf itself. Because of this, fresh leaves need to be collected once a day, as once they dry out, they are no longer suitable as a food. After a few days, the larvae will be capable of eating mature leaves. Watch closely to make sure there are enough leaves for all the caterpillars. After the first few days, the larvae will have grown too big to fit in their Petri dish and will need to be moved to a larger container. They can be kept in anything ranging from a plastic bin to a cardboard box. It's not necessary to have a lid, as the larvae will not crawl away – they will just stick to the mulberry leaves. Having a lid can keep the leaves from drying out, however, and reduce the frequency at which the silkworms need to be fed. If you do have a lid on the box, make sure it has holes so that the larvae stay well ventilated. The container should be kept in a shaded location and be protected from predators such as birds. It's important to remove eaten leaves from the bottom of the container, as when they accumulate, they can begin to grow mold. Dried leaves, as well as "frass", or the silkworm droppings that collect in the bottom of the bin will make excellent amendments to a compost pile.

Over the next few weeks the silkworms will grow rapidly. By the time they are two inches in length, their bodies contain mostly silk and they lose their value as a livestock feed. Before they reach that length, most of them should be fed to chickens or fish. You can keep your silkworm population going by allowing at least twenty of them to spin cocoons and hatch into adults (Figure 7.3). They will mate and lay eggs, which can be collected and used to spawn the next generation of worms. If you are running into the end of the growing season, you can keep the eggs in the refrigerator over the winter and hatch them in the spring when the leaves have returned to the trees (mulberry leaves can be kept fresh for a couple days by harvesting small branches and putting them in water – this is useful if you don't live close enough to a tree that can be visited on a daily basis).

Maple trees

In the late winter when daytime temperatures push above freezing – yet still fall below the freeze point in the night – and the number of daylight hours start to increase, the sap of maple trees (genus *Acer*) starts to flow from its roots to its buds as it prepares for the coming spring. The environmental triggers of temperature and light entice the tree to convert starches (stored throughout the winter in its roots) into sugars, which are carried in liquid sap upwards through the trees' vascular system. By creating small holes in the trees' bark, the sugary sap (a 2-percent sugar content) can be collected and boiled down into maple syrup, a deliciously sweet product that is now known and sold worldwide. When employing best practices, maple trees can be sustainably tapped year after year without negatively impacting the health of trees (Berg, 2012). In this sense, maple tapping is regarded as a sustainable non-timber use of tree products and has parallels

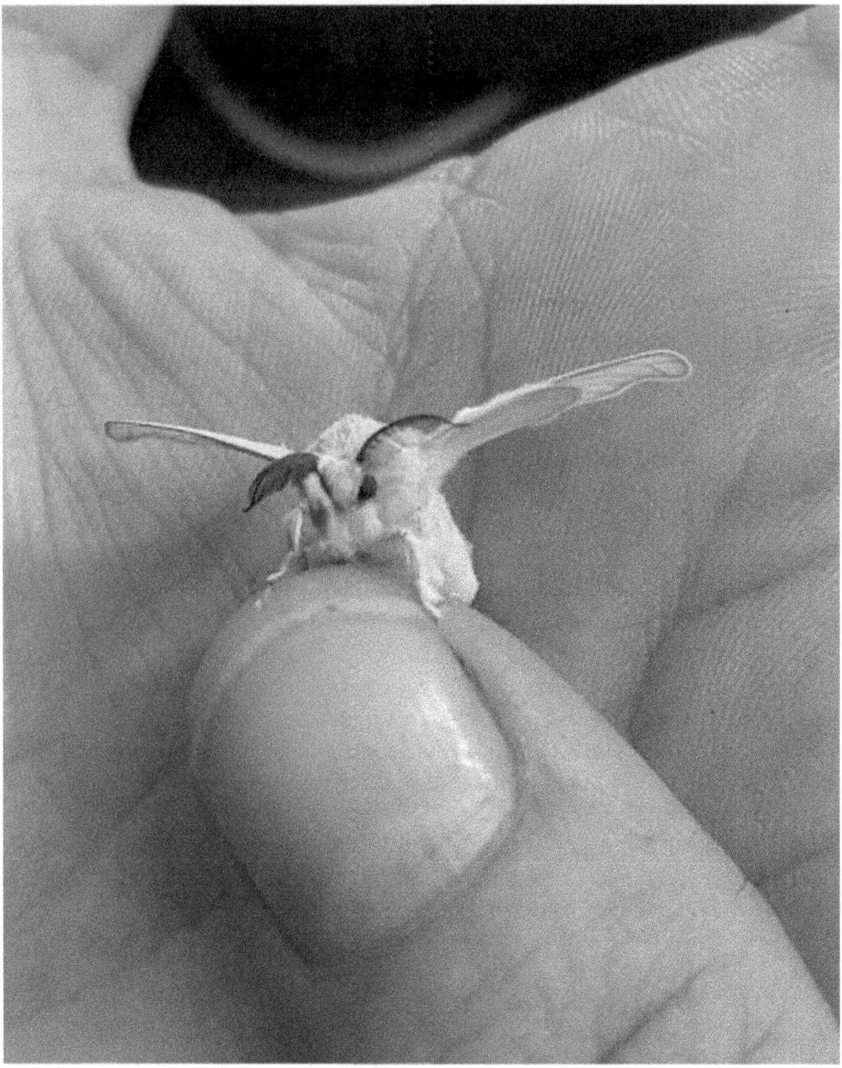

Figure 7.3 Silk Moth
Source: Scott Kellogg

with similar practices throughout the world including rubber tapping, pulque, and palm wine production (Osa, 2013). The flow of sap is one the first signs of spring in the region – connection with it gives a feeling of movement, life, and animacy during a time of year that may otherwise feel endlessly bleak. While traditionally associated closely with rural life and spaces, maple tapping can be carried out in urban environments as well with little modification. A fun and multisensory activity, tapping urban maple trees is an excellent way to foster a sense of reciprocity, biocultural diversity, and urban forest justice (Poe, 2013) among participating youth.

Maple tapping is a technology originally developed by indigenous peoples of eastern North America for whom it had great nutritional and cultural significance (Kimmerer, 2013) and is still practiced by their descendants today (Murphy, 2009). The practice was thereafter adopted by European settlers in the northeastern United States and Canada (for a time maple sugar was regarded by Northern abolitionists as a morally preferable, locally made alternative to cane sugar produced through slave labor (Sturges, 2018)). While traditionally carried out by hanging hand-collected buckets from taps or "spiles" in the sides of trees (Figure 7.4) and boiling off water with locally harvested firewood, maple extraction in the later twentieth century has expanded into a large-scale industry that produces millions of gallons annually. Employing yield-increasing technologies such as reverse osmosis and vacuum pumps, industrialized sugarbushes consist of miles of tubing crisscrossing the forest that deliver sap to fossil fuel-powered evaporators. More recently, a high-yield method has been developed that involves cutting the tops off sapling maples that, when coupled with sophisticated technology, effectively transforms maple tapping from a perennial semi-wild practice into a highly controlled annual monoculture (Brown, 2013). The high capital costs of these innovations have made it extremely difficult for small-scale maple producers to remain competitive (Gorelick, 2016). The focus of this section will be upon more traditional technologies and practices requiring reduced capital and energetic expenses.

Within their North American range (primarily the northern and eastern half of the country, extending into southern Canada), maple trees form dense stands and are often the dominant tree species in particular ecosystems (Duchesne, 2005). The trees are abundant inside of cities within this extent as well, with maples consisting of 80 percent of all street trees in Worcester, Mass. (The trees were planted extensively in the 1950s on account of their availability and shading, however, they are now threatened by the Asian long-horned beetle) (Green, 2008). While the sugar maple (*Acer saccharum*) is regarded as producing the best quantity and quality of syrup, it is possible to produce fine syrup from any tree in the Acer genus including red maples, silver maples, box elders, and the Norway maple (other tree species, such as sycamores, birches, and walnuts are also tappable but will result in low yields) (Farrell, 2013).

At the Radix Center, maple syrup production is an integral component of our urban ecosystem justice curriculum. Each year we are granted a permit from the City of Albany to tap the stands of maple trees in neighboring Lincoln Park, as well as a number of trees on public school campuses. Youths are engaged in every aspect of the process including selecting trees to be tapped, drilling and tapping, regular sap collection, and sap boiling/syrup finishing. The finished maple syrup (which does not legally need to be produced in a commercial kitchen – unlike jams, jellies, and other value-added products) can then be sold by the youth throughout the year.[9]

Figure 7.4 Tapped Maple Trees

Source: photo by Cait Denny

Making maple syrup

One fun aspect of maple syrup production is the wide range of scales at which it may be practiced, ranging from the aforementioned industrial level to the backyard micro-scale. Entry-level syrup production can begin with materials as simple as a one-gallon milk jug and a pot on a stove top. As with most things, I highly recommend starting a maple syrup production program on a small level and building up incrementally following initial success. Producing finished syrup, especially in its final stages, is a precise process that can easily result in stove-top spillovers and burnt char messes.

What you will need:

- Access to one or more maple trees
- Drill and 7/16" drill bit
- Maple tree tap, or spile (typically made of metal but can be made from wood or plastic)
- Collection bucket (could be anything from a five-gallon bucket to a one-gallon jug)

To begin, locate a maple tree. As mentioned earlier, sugar maples are the most preferred species, however, others are acceptable. Trees are most easily identified while their leaves are still out, otherwise, you'll be relying on bark identification. When temperatures are above freezing in the daytime, but still below freezing at night, drill a one-inch deep hole at a slight upward angle (so that sap will flow down) into the side of the tree at approximately chest height. Holes are best made on the south-facing side of the tree, which is more likely to warm and flow faster. If sap is flowing and the tree is healthy, you should notice sap coming out of the hole right away. Allow the sap to clear any wood dust and then insert your spile into the hole – tap it in gently with a hammer or rock so that it does not fall out. At this point, you are ready to begin collecting sap. The simplest way to do this is to make a small hole in the side top of a one-gallon plastic milk jug – push the jug over the end of the spile and allow it to hang from it. Once the jug is full of sap (the time will vary depending on flow conditions), you are ready to boil it down into syrup. Remember that at best there is a 40:1 ratio of sap to syrup, so a single gallon of sap will yield very little syrup. You may wish to transfer the sap into a larger container, keeping it refrigerated until you've collected a greater amount to be boiled down at once.

Boiling with small sap quantities is most easily done over a stove top where you can be precise with heat controls. Pour sap into a pot or pan (the shallower and wider the better – maximum surface area is conducive to evaporation) and bring to a boil (Figure 7.5). As water evaporates, the sap will become browner and sweeter, eventually turning into syrup. Using a syrup hydrometer, it is possible to accurately determine when sap has turned to syrup, otherwise you will need to rely on your senses of taste and texture. If the syrup boils down too low, it will quickly spill over the top of the pot and create a horrible mess or burn completely. It is

Figure 7.5 Boiling the Sap

Source: Scott Kellogg

therefore critically important to watch it closely in its final stages and be ready to shut off the heat the second it is ready. When the syrup is finished, pour it into sterilized jars and enjoy. Syrup has a very long shelf life when properly canned and sealed and after opening can be kept for long periods in a refrigerator.

Scaling up

Once you have mastered small-scale syrup making, you could consider scaling up your production. At Radix, we put out approximately fifty five-gallon buckets and taps to collect sap. Large buckets are too heavy to be hung from spiles when full, so they are placed on the ground and a length of sap tubing is run from the spile through a tight-fitting hole in the lid of the bucket (buckets must be kept sealed in order to keep out rain). Sap is collected on a regular basis and transferred to a large storage barrel/pot that is kept cool by being packed in snow (it is important to keep sap cool – if it gets too warm it becomes vulnerable to bacterial contamination/fermentation).

Black soldier flies

Another insect within the "zone of reciprocity" that can be partnered with for purposes of both protein production and waste management is the black soldier fly. Black soldier flies (*Hermetia illucens*), or BSF, are a species of fly in the family Stratiomyidae that superficially resemble a wasp more than a fly. They have beautifully iridescent wings, stubby antennae, and streamlined bodies (Figure 7.6). Slow flyers incapable of biting or stinging, they can be picked up and handled without difficulty, often landing on humans while resting. As adults, they do not eat – all their feeding is done while in their larval states. Since adults have no interest in food, they do not buzz around picnic tables or come into kitchens and spread disease in the way that common houseflies do (BSF larvae actively compete with house fly maggots, further reducing their pestilence). Upon metamorphizing into an adult, BSF only live for a few days – long enough to reproduce, lay eggs, then die shortly afterward. A short, elegant cycle of living and dying.

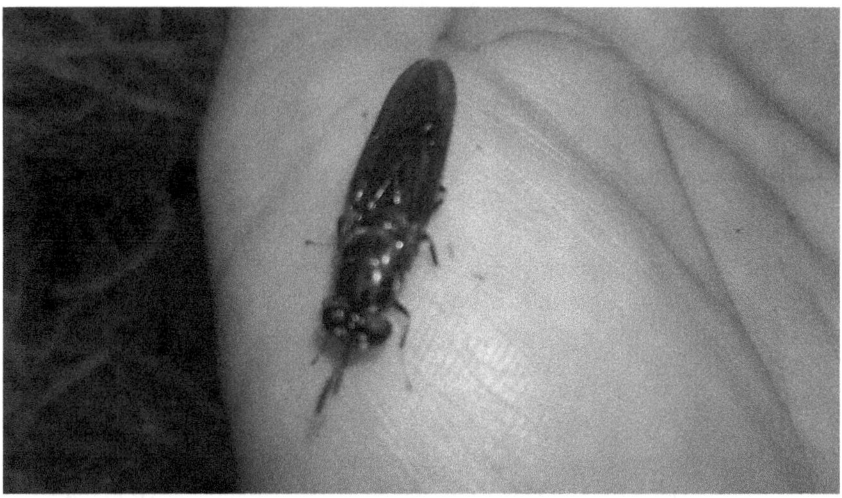

Figure 7.6 Black Soldier Fly

Source: Scott Kellogg

While in their larval phase, BSF are voracious consumers of organic wastes, particularly "putrescent" wastes including meat, dairy, manures, carcasses, and cheese – high-nutrient materials prone to putrefaction, decay, and odors that are otherwise difficult to compost with conventional methods. Despite being historically regarded as a pest (Tomberlin, 2020), the ability of black soldier flies to rapidly consume putrescent wastes has now been widely recognized and studied and put into use throughout the world (Nguyen, 2015).[10] BSF composting can be carried out on scales ranging from the mega-industrial to the micro-sized DIY and can be either rigorously controlled or minimally managed (Nana, 2018). This scalability and accessibility makes BSF composting a promising tool of both urban and rural communities for ecological regeneration, particularly for upcycling putrescent wastes into fertilizer and feedstock.

In addition to their waste reduction capabilities, BSF have also been heavily studied for their use as a livestock feed. Rich in both fat and protein, BSF larvae will be gleefully eaten by chickens, reptiles, and notably, fish – therein potentially helping to reduce the aquaculture industry's use of bycatch-based feeds (a significant cause of oceanic depletion and obstacle to the development of sustainable fish farming) (Swinscoe, 2019). While significant legal and cultural boundaries remain, there is also considerable interest in growing BSF as a protein source for humans (Wang, 2017). Not without precedent, BSF entomophagy remains a practice of some traditional societies (Chung, 2010).[11] Topping off their list of benefits are the facts that BSF wastes, a dark-crumbly compost, are themselves excellent gardens fertilizers (Sarpong, 2019) and that the fatty oils in BSF may be extruded and used as a fuel for bio-powered maggot-mobiles (Li, 2011). By upcycling waste products into an edible protein, BSF have promising potential to create more cyclical urban metabolisms in regard to both waste and food.

BSF are native to the Americas, although they can now be found throughout many tropical and temperate regions of the world, having been either intentionally or accidentally introduced.[12] As adults are most active when temperatures are above 80 degrees Fahrenheit, low temperatures have limited their northward migration (Spranghers, 2017).[13] Due to climate change, however, the range of the BSF will surely increase, along with that of many other insect species (Bradshaw, 2016). Without any solid data to support it, I have personally observed BSF in northern locales with greater frequency over the past decade. I suspect that their larvae are surviving winters inside of outdoor compost piles, which provide both heat and food source (Figure 7.7). It is possible that BSF occurrence in cities is greater than surrounding rural areas, with the urban heat island and abundance of organic waste giving them an edge on survival as it does similarly for other insects (Frank, 2020).

Relationships

Despite many years of trying, I have never been reliably successful in controlling the breeding and growth of the black soldier fly. My attempts at containing and intensively managing their life cycle have been hit or (mostly) miss, with successes

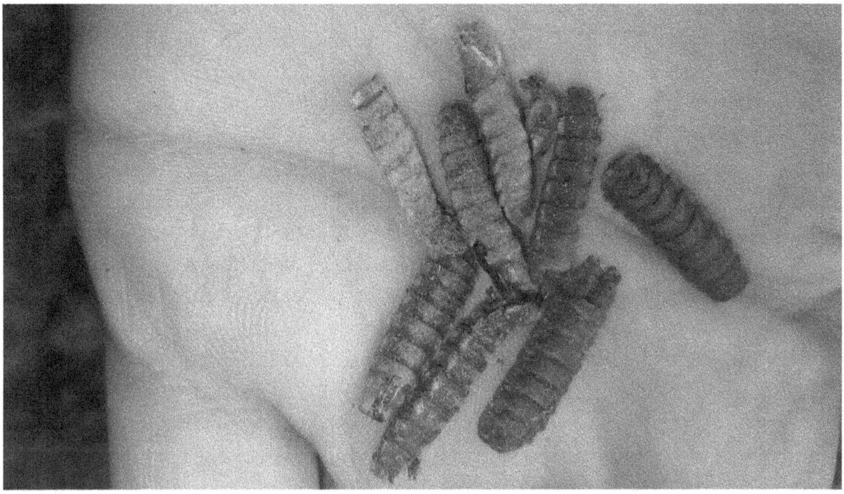

Figure 7.7 Soldier Fly Larvae

Source: Scott Kellogg

often being accidental and short-lived. A dense colony of BSF raised in confinement will frequently collapse for no apparent reason, some aspect of humidity, temperature, food, pathology, or other unknown variable causing them to tailspin into a positive feedback loop of mass sudden die-off. This doesn't mean I have not had robust BSF cultures on our land, they have just been in open-air compost piles under minimally managed conditions, living, reproducing, and dying on their own terms. I provide them with food waste and in return they rapidly consume it, giving me finished compost as a by-product (along with the ability to occasionally harvest some of their larvae as animal feed). This resistance to total domestication and regulation is partly what makes BSF fascinating to me – they thrive under open, chaotic, semi-wild conditions, willing to engage in reciprocal ecological relationships, but with their own agency.

BSF's resistance to being contained does not mean that it is impossible. On the contrary, there are industrial-scale BSF breeding facilities throughout the world that seek to reduce, mechanize, and standardize the fly's life cycle so that it may be carried out continuously while in confinement without regard for outside environmental conditions or seasonal fluctuations and producing neither odors nor offense – an entirely regulated biological waste consumption machine fully compatible with the aesthetic demands of modernity. Environmental anthropologist Amy Zhang brilliantly explores this "circularity and enclosure" in an ethnography of a Chinese facility tasked with mending a city's broken ecological waste metabolism with BSF while crafting an illusion of seamless techno-utopianism (Zhang, 2020).

Cultivation of BSF can happen on a micro-scale as well – there are multiple online DIY designs for BSF composters that use little more than a five-gallon bucket and some tubing, as well as several premanufactured molded plastic BSF composters that are commercially available. They all feature a container where BSF larvae are kept and where food waste is deposited for them to eat. The designs take advantage of the "self-harvesting" behavior of BSF – upon pupation, the larvae will climb up a ramp to a drier, higher spot where it is easier to fly away from after becoming a winged adult. The deceptive ramp terminates in a hole through which the BSF fall – into a collection bucket where they are then gathered. Unlike houseflies, BSF do not lay their eggs directly into food but instead in a nearby location. When the eggs hatch, the baby larvae will crawl to the food source. The designs also take advantage of this trait, allowing the adults to deposit their eggs on the outside of the container but leaving small openings for the larvae to crawl through (Figure 7.8).

It may be possible to keep a colony alive through the winter in an enclosed BSF composter. Soldier fly larvae produce a great deal of heat through their own metabolic processes. Provided they are sufficiently fed, this heat can be enough to keep the larvae alive and active. Although they will not be hatching into adults until the warm weather returns, the larval colony can still be kept alive and continue to process putrescent wastes. Raising BSF in cold climates is an active field of research (Alvarez, 2012).

The alternative to raising BSF in a designated composter is to instead cultivate them in a semi-wild manner where they can effectively reduce wastes without continual maintenance on your behalf (it will also be less convenient to harvest the larvae). The first step in building a reciprocal relationship with BSF is to establish a healthy local population. Compost piles make ideal sites for BSF colonization, providing both food and warmth to the larvae and offering the possibility of overwintering a population. If BSF are already abundant in your area, they will likely arrive in a compost pile on their own. If they are not already common, or if you want to speed up the rate of colonization, another strategy is to "seed" a local population. This can be done by purchasing a container of black soldier flies, commonly sold as a reptile food under various trade names. While the greater the number of larvae you initially inoculate a compost pile with will determine the likelihood of success, between 250 and 500 larvae is a good start. Once a population is established, a colony can grow at a nearly exponential rate, as the female BSF (which are attracted to pheromones given off by the larvae) will lay upwards of 500 eggs each (Givens, 2013). While compost piles dominated by BSF can have putrescent foods put in them, be sure all the food waste is being eaten before adding more so as not to create odor problems. Larvae can be easily harvested from pile, as they typically congregate at its top. Pushing aside the top few inches of material will reveal them, after which they can be collected with a small shovel or a spoon. Provided the compost pile is large and fresh enough, it may continue to produce enough heat through the winter to keep the population of BSF alive. During the cold season, the larvae may migrate to the core of the pile where it is the warmest and may not be visible near its surface.

Figure 7.8 Insectorium

Source: Scott Kellogg

Educational applications

It would be hard to describe black soldier fly larvae as being beautiful – admittedly the first time I ever saw them, they filled me with feelings of disgust – super-sized maggots crawling throughout food waste slime in a compost pile. For this reason, they will likely trigger "disgust sensitivity" (Jensen, 2019) within most people, particularly among those unfamiliar with the composting process.

Honeybees and native bees

Honeybees are a type of eusocial (Gowdy, 2013) insect that has been kept, tended, or cultivated by humans for their honey and other products for over seven thousand years (Oldroyd, 2012). While the western honeybee (*Apis mellifera*) is the species most raised today, there are traditions of harvesting honey from several bee species throughout the world (including *Melipona beecheii*, a stingless bee which was raised by the Mayans in the Yucatan (Villanueva-G, 2005)). The western honeybee, originally native to Europe, Africa, and the Middle East (Han, 2012), has now been spread by humans throughout the globe with an estimated ninety million kept beehives worldwide.

Industrial monocultural agriculture is highly dependent on *Apis mellifera*, relying on them nearly exclusively to pollinate a great number of flowering staple food crops including almonds, apples, blueberries, melons, broccoli, and others. Insufficient insect pollination results in low or no fruit set and cannot be substituted by technological means (despite dangerous technofix attempts such as robot bees (Potts, 2018)). Making a migratory loop throughout the country (a process that is highly stressful to bees (Simone-Finstrom, 2016)), commercial beekeepers transport hives via truck bed to different farms as their crops start to bloom and are an integral part of the world's commercial food system. While much concern has been raised about Colony Collapse Disorder (CCD), a still-mysterious and multicausal affliction that caused significant losses of beehives in the early 2000s (Andrews, 2019), CCD poses more of a threat toward industrialized food production than to insect pollinators' populations in general (although many of CCDs causes – pesticides and habitat destruction – are also contributing to the broader losses of "the insect apocalypse" (Goulson, 2019), and honeybees populations serve as a proxy for insect health at large).

Although modern-day beekeeping is a highly managed practice that can involve genetic manipulation and the use of pharmaceuticals and pesticides, I still situate honeybees in a semi-wild, reciprocal zone category that is resistant to both domestication and domination. Even in the most chemically dependent and intensively monitored commercial beekeeping operations, there is a recognition of the bees' own powerful agency in their relationships with humans along with a sense of collaboration versus control that distinguishes bees from other livestock such as chickens or cows (Nimmo, 2015). For one, honeybees defy containment and property boundaries – as honeybees have a typical foraging range of two to three miles (Beekman, 2000), no fence, netting, or structure can be built to confine them.

Like the pigeons at Radix, bees are transgressive agents – freely leaving the site and providing community pollinating services to neighborhood gardens abound. When a colony has grown too large or if conditions are less than ideal, the hive can abscond outright in full or in part by "swarming" – leaving their human manufactured boxes and taking up residence elsewhere in a tree, wall cavity, or other hollow structure. Visible as a dark and writhing mass of bees clinging to a tree branch, "wild swarms" originating from these feral colonies can with skill and luck in turn be "captured" and relocated to a bee box, providing a beekeeper with a free colony of local winter-hardy bees (Figure 7.9). In the springtime, when bees typically swarm, I often receive calls from panicked residents notifying me of a swarm in need of removal. Assuming they are dangerous, few realize that bees are exceptionally docile when swarming. Despite multiple attempts, I have yet still to be successful in collecting a wild swarm, a reminder to me of bees' own wild agency and fiercely independent spirit.

Raising honeybees in urban environments has become more popular in recent years (Moore, 2013), with beekeepers maintaining hives in underutilized spaces such as vacant parcels, parking lots, and rooftops. Honeybees are kept at Radix primarily for their service as pollinators (we allow them to keep the honey they produce to improve their chances of surviving the winter). With thousands of young to feed and a willingness to forage from a wide range of plants, honeybees are highly effective pollinators, greatly boosting the yields of fruit trees and vegetables alike. They also are fantastic for educational purposes – it is great for children to taste raw honey straight from a comb or to handle a non-stinging drone (male) bee (Figure 7.10).

Figure 7.9 Honeybees at Radix

Source: Scott Kellogg

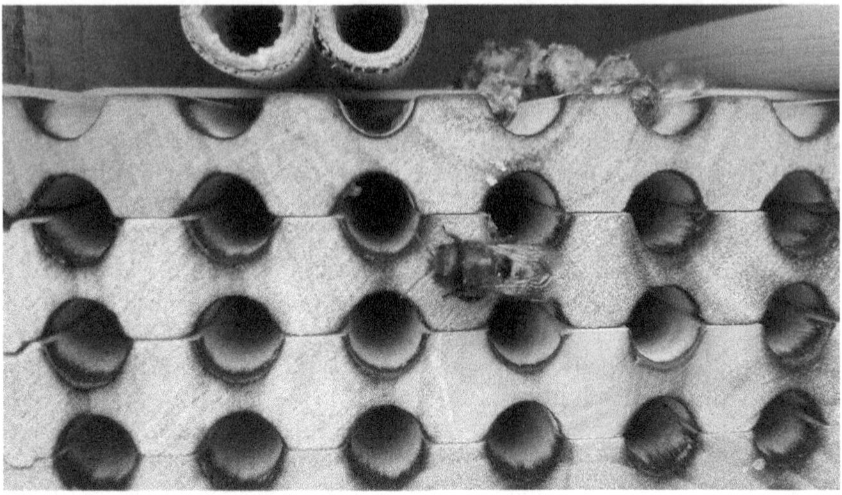

Figure 7.10 Mason Bees Emerging from Bee Hotel
Source: Scott Kellogg

It is made clear to visitors, however, that honeybee pollination is a parallel strategy to supporting native pollinating bees. Recognizing the challenges of beekeeping (CCD, varroa mites, nosema disease) and that *Apis Mellifera* is not native to North America, we simultaneously create habitat and food for the hundreds of species pollinating bees, wasps, butterflies, and other insects native to New York (Cornell, 2020). While not honey producing, when present in sufficient numbers these native insects can provide pollination services that are equal to (Winfree, 2007) or potentially better than (MacInnis, 2019) honeybees. For certain pollinators, such as mason bees, we will construct housing and nesting tubes called "insect hotels" to enhance their productivity (Fortel, 2016). Our main strategy for supporting native insects, however, is to allow spontaneous vegetation to arise and go to flower in the marginal spaces alongside of gardens, creating un-mowed urban meadows (Norton, 2019). Pollinators drawn to the flowers of the wild plants will also be attracted to the blossoms of the nearby vegetables and fruit trees, therein creating food security and biodiversity enhancement simultaneously. In this way, it is possible to illustrate biocultural diversity, demonstrating to urban youth clear linkages of how the well-being of insects is interwoven with food production and in turn their personal, family, and community's health.

Wine cap stropharia

The last example of semi-wild relationships I wish to describe, if only briefly, is the cultivation of the wine cap stropharia (*Stropharia rugosoannulata*). Wine caps are a gourmet edible type of litter-decomposing fungi that thrive in non-sterile environments, growing rapidly throughout substrates such as soil and wood chips. At the Radix Center, wine caps are grown in the chipped walkways between garden beds and the mulch donuts around the bases of fruit trees. When conditions are right (typically spring and fall following a rain) the mushrooms will fruit and grow to enormous sizes approaching ten inches in diameter (thus the reason for their moniker "garden giants"). The mycelial lenses, or clusters of wood chips tightly bound with fungal hyphae, can be transplanted into new areas to start new wine cap colonies. Provided they are "fed" new wood chips on an annual basis, stropharia patches will reliably produce each year. I include this fungal species within the zone of reciprocity on account of its predilection for non-sterile habitats. Many other edible fungi species demand highly specific growing substrates and conditions and must often be initially cultured in sterile laboratory conditions with technical expertise and significant capital costs. Furthermore, sterilizing or pasteurizing the grain or straw substrates needed by these fungal strains is time and energy intensive. The technical and financial requirements of these fungi limit the number of people on global scale who may be able to cultivate them. The fact that wine caps may be grown in literally dirty conditions, and then may be propagated and shared through simple division, greatly increases the potential for them to be grown on a broad scale. While their fruiting times may be somewhat irregular and defy attempts at precise control, they nonetheless can provide a fairly reliable protein source to their human counterparts, placing them in the category of "semi-wild".

Beyond being a food source, research has revealed the potential of using wine caps for mycoremediation of a number of organic pollutants that contaminate urban soils (see the Urban Soil Justice chapter for more details) (Pozdnyakova, 2018); (Valentín, 2013); (Castellet-Rovira, 2018). Their ability to rapidly spread through the soils of polluted vacant parcels while oozing enzymes makes them a promising candidate for remediating toxins in the heterogynous, uncontrolled, wild, and dirty environments that are urban brownfields (Figure 7.11).

Conclusion – zoonoses and the limits of nature–society reunification

This chapter has demonstrated the relationships and pathways between humans and the semi-wild that attempt to bridge the duality of nature and society, a false dichotomy that I believe lies at the core of many of the attitudes and behaviors that drive ecological degradation and human exploitation. An emerging area of study (that is underexplored in this chapter) is around the idea of urban microbial rewilding or designing urban spaces to maximize their microbial diversity for public health benefit (Flies, 2018); (Mills, 2017). This is a fascinating and important field of research as studies have suggested that many autoimmune disorders may be attributed in part to a lack of exposure to beneficial microorganisms in early childhood (Wasko, 2020); (Panelli et al., 2020). Are there, however, potential negative consequences that might result from the haphazard mixing of species and their associated microbiomes? Even before the creation of the nature–human dualism, were there still some codes of respect, distance, and balance that

Figure 7.11 Wine Cap Stropharia

Source: photo by Ben Atwood

maintained the discordant harmony between human and non-human worlds that must now be remembered? Despite these codes, did unintentional crossovers still occur that created disruptions? Zoonoses, diseases with origins in non-human animals (chiefly viruses and bacteria), are a reminder and manifestation of what potentially can occur when species overlap and combine in novel configurations.

The earth's virosphere is vast, with estimates of potentially ten trillion species yet to be discovered (Zimmer, 2020). The vast majority of those viruses coexist with humans non-threateningly, with viral endosymbiosis being a major driver of evolutionary transformation (Margulis, 2008). Occasionally, however, a virus will jump from one species to another and cause illness in its efforts to coevolve (generally to a less virulent strain). The human consequences of these zoonotic illnesses can be devastating, with the histories of bubonic plague, smallpox, measles, certain influenzas, HIV, rabies, Lyme disease, and Ebola illustrating the potential effects. Many of these zoonoses are the result of agricultural intensification, animal domestication, and ecological disturbance (Jones, 2013) – when species are driven out of their normal habitat or when they are newly hunted, opportunities are created for viral spillover (Quammen, 2012).

As I write this, we are in the midst of the coronavirus pandemic, another zoonotic virus originating in bats that has crossed over to humans, taking over a million human lives so far (Mega, 2020) and wreaking havoc upon the world. Coronavirus is very much a disease of the Anthropocene, with ecological disturbance, social inequality, hyper-connectivity, and political ineptitude compounding to amplify the suffering created by it (O'Callaghan-Gordo, 2020). In this way, COVID-19 has served as a grim reminder of our planetary interconnectivity, exposing the myths of individual or national isolation and separation for the dangerous illusions that they are.

With so much effort now to control the spread of the disease through its focus on vaccines, contact tracing, masks, and so on, the public health discourse frames the coronavirus as a uniquely human challenge. While appreciating this urgency, and acknowledging that good work has been done in mapping the socioeconomic and environmental inequities of the disease (Bowleg, 2020); (Lenzen, 2020), I would hope that a parallel discourse could be undertaken that frames COVID-19 and zoonoses in general as a more-than-human phenomenon, planning prevention of future pandemics with the most holistic strategies. Specifically, I am eager to avoid an overly technocratic response to future disease threats that creates greater ecological alienation, further separating nature from society and placing blame upon the world's poorest. Such holistic frameworks are already being developed, with the platforms of one-health, eco-health, and planetary health providing global systems-level analyses of the interrelationship of human and ecological health (Lerner, 2017). Further philosophical work needs to be undertaken that frames pandemic response as a post-normal science embracing complexity and that re-narrativizes the human–non-human relationship (Waltner-Toews, 2017). In this analysis, concerns about social justice, power, politics, and global equity must also remain central, resisting the tendency of affluent researchers in the Global North to place the responsibility for causing and preventing disease upon subsistence farmers and hunters (Asayama, 2020); (Scoones et al., 2018).

I'm eager to avoid policies enacted to prevent future pandemics that favor practices of agricultural modernization that would further erode traditional lifeways. For instance, there has already been a push to confine chickens to darkened and cramped enclosures to prevent any possibility of them mixing with wild bird populations and being exposed to avian flu (Davis, 2006). Along with the forced preventative culling of backyard chicken flocks, such actions were taken in Cairo, Egypt in 2005 under the logic of "biosecurity" recommended by the United Nation's Food and Agriculture Organization (Hinchliffe, 2008). These measures unquestionably increase the profits of a handful of centralized industries and destroy a critical component of food security for the urban poor. While the threat of disease is serious, I believe it is important to implement more measured approaches that:

- Emphasize best practices of health and safety for small-scale poultry flocks (recognizing that birds that are well fed, non-stressed, spaced apart, genetically diverse, and exposed to sunlight are less likely to get sick than their inbred counterparts in confined operations)
- Question the dynamics of power and access as they relate to who controls the food supply
- Recognize that cities are multispecies environments where interspecies entanglement is inherent

(Nading, 2013)

Keeping these principles in mind, and in anticipation of the authoritarian technocratic policies potentially to be rushed in following the end of the coronavirus pandemic, I believe it is time to reexamine the "politics of conviviality" (Hinchliffe, 2006) that describes the "recombinant ecologies" of cities and the multispecies coevolutionary assemblages they are. Let us hope that biosecurity mindsets driven by fear and the impulse to control do not further modernize urban ecosystems by segregating and exterminating species but rather foster relationships of reciprocity and respect that recognize the enormous fluidity and interchange that occurs between species and their exchanging virospheres. It is my hope that the relationships outlined in this chapter may provide a few examples of how just and reciprocal urban biocultural diversity may be cultivated.

Notes

1 See concepts like "biophilic cities".
2 The exact causes of the luxury effect are debated – it may result from wealthier residents' preferences for more biodiverse areas or be a consequence of the large "cultured diversity" of cultivated ornamental plants found in wealthier neighborhoods. In my own experience and that of others. I have noticed multiple exceptions to the luxury effect where low-income neighborhoods seem to have higher rates of biodiversity on account of the preponderance of unmaintained and forested lots/parks/cemeteries that unintentionally function as biodiversity refuges.
3 I will emphasize by restating that this is indeed a continuum – in no way intends to reify the false dualism between the domestic and the wild. Rather, I see "domestic" and "wild" as vague poll points of reference along a spectrum of possible relationships. Many times, species will easily move from one poll toward another. "Feral" animals are a good example of this – once domesticated species or their descendants becoming less co-dependent on humans. Similarly, formerly wild species can become synanthropic – tolerant of humans and successful at cohabitation with them.

4 Later in history, pigeon manure was used for both tanning leathers and for manufacturing saltpeter, an additive in gunpowder. Additionally there are stories of British and French armies fueling their war machines with the huge guano piles commandeered from the pigeon coops of wealthy landowners.
5 It should be mentioned, though, that pigeons are definitely at risk from aerial predators, such as falcons and hawks. Many birds of prey, in fact, choose to nest in skyscrapers, attracted to cities by the abundant pigeon food supply. There's not much that can be done to protect free range pigeons from raptors. You just have to hope that, considering how many pigeons live in cities, statistically your birds are less likely to be hunted.
6 While *Bombyx mori* is the most commonly raised species of silkworm, there are other domestic and semi-wild species of silkworm that eat other species of plants besides mulberries.
7 It should be noted that the global silk industry has often relied upon bonded child labor (especially in India) in order to produce its products – essentially a form of slavery with terrible working conditions for children.
8 Florence, Massachusetts is one example. The nineteenth-century utopian community "Northampton Association of Education and Industry" (of which Sojourner Truth was a member) once ran a cooperative silk industry there.
9 Before the project began, we were concerned about the potential for lead or other contaminants to be soluble in the liquid sap and to concentrate in the finished syrup. After being unsuccessful in finding any published studies regarding the safety of urban syrups, we partnered with researchers at Siena College to run an analysis of the finished product. Their testing with an X-ray fluorescence machine revealed concentrations of metals in the Albany syrup that were comparable to those found in a control sample taken from a rural location.
10 The internet is full of time-lapse videos displaying soldier fly's ability to reduce various animal carcasses to bones in matters of hours.
11 Care should be always be taken to be certain that BSF are dead before being consumed as they are capable of otherwise surviving in the human digestive track, resulting in the horrific condition of myiasis.
12 The potential impact of black soldier flies on ecosystems to which they are not native is understudied, however, at least one study has rung the alarm bell following their apparent escape from a breeding facility.
13 Inaturalist data shows BSF ranging into the northeastern United States and parts of northern Europe.

Works cited

Akkerman, S. F. (2011). Boundary crossing and boundary objects. *Review of Educational Research, 81*(2), 132–169.

Alberti, M. (2008). *Advances in urban ecology integrating humans and ecological processes in urban ecosystems*. New York: Springer Science+Business Media.

Alvarez, L. (2012). *The role of black soldier fly, Hermetia illucens (L.) (Diptera: Stratiomyidae) in sustainable waste management in Northern Climates*. Thesis, University of Windsor.

Andrews, E. (2019). To save the bees or not to save the bees: Honey bee health in the Anthropocene. *Agriculture and Human Values, 36*(4), 891–902.

Asayama, S. e. (2020). Are we ignoring a black elephant in the Anthropocene? Climate change and global pandemic as the crisis in health and equality. *Sustainability Science*, 1–7.

Barber, E. J. (1991). *Prehistoric textiles: The development of cloth in the Neolithic and Bronze Ages with special reference to the Aegean*. Princeton, NJ: Princeton University Press.

Beatley, T. (2011). *Biophilic cities: Integrating nature into urban design and planning*. Washington, DC: Island Press.

Beekman, M. a. (2000). Long-range foraging by the honey-bee, *Apis mellifera* L. *Functional Ecology*, *14*(4), 490–496.

Bennett, C. E. (2018). The broiler chicken as a signal of a human reconfigured biosphere. *Royal Society Open Science*, *5*(12), 180325.

Berg, A. v. (2012). *Defining modern, sustainable tapping guidelines*. Burlington, VT: University of Vermont Proctor Maple Research Center.

Berger, J. (2013, April 3). Breeding pigeons on rooftops, and crossing racial lines. *The New York Times*.

Bowleg, L. (2020, July). We're not all in this together: On COVID-19, intersectionality, and structural inequality. *American Journal of Public Health*, 917.

Bradshaw, C. J. (2016). Massive yet grossly underestimated global costs of invasive insects. *Nature Communications*, 1–8.

Brown, J. E. (2013). *Remaking maple: New method may revolutionize maple syrup industry*. University Communications, University of Vermont. Retrieved from www.uvm.edu/uvmnews/news/remaking-maple

Buizer, M. B. (2016). Governing cities reflexively: The biocultural diversity concept as an alternative to ecosystem services. *Environmental Science & Policy*, *62*, 7–13.

Cai, F. a. (2016). Seasons and neighborhoods of high lead toxicity in New York City: The feral pigeon as a bioindicator. *Chemosphere*, *161*, 274–279.

Castellet-Rovira, F. e. (2018). *Stropharia rugosoannulata* and *Gymnopilus luteofolius*: Promising fungal species for pharmaceutical biodegradation in contaminated water. *Journal of Environmental Management*, *207*, 396–404.

Chung, A. Y. (2010). *Edible insects and entomophagy in Borneo: Forest insects as food: Humans bite back*. United Nations, NY: Food and Agricultural Organization of the United Nations.

Clark, C. (2019). *The communitarian moment: The radical challenge of the Northampton Association*. Ithaca, NY: Cornell University Press.

Cooke, A. O. (1920). *A book of dovecotes*. London: TN Foulis.

Cornell, U. (2020). *Bee diversity in New York*. Retrieved from Pollinator Network at Cornell, https://pollinator.cals.cornell.edu/wild-bees-new-york/bee-diversity-new-york/

Davis, M. (2006). *The monster at our door: The global threat of avian flu*. New York: Macmillan.

Duchesne, L. e. (2005). Changes in structure and composition of maple – beech stands following sugar maple decline in Québec, Canada. *Forest Ecology and Management*, *208*(1–3), 223–236.

Elmqvist, T. e. (2013). *Urbanization, biodiversity and ecosystem services: Challenges and opportunities: a global assessment*. Dordrecht: Springer.

Farrell, M. (2013). *The sugarmaker's companion: An integrated approach to producing syrup from maple, birch, and walnut trees*. White River Junction, VT: Chelsea Green Publishing.

Flies, E. J. (2018). Cities, biodiversity and health: We need healthy urban microbiome initiatives. *Cities & Health*, *2*(2), 143–150.

Fortel, L. e. (2016). Use of human-made nesting structures by wild bees in an urban environment. *Journal of Insect Conservation*, *20*(2), 239–253.

Frank, S. D. (2020). Can cities activate sleeper species and predict future forest pests? A case study of scale insects. *Insects*, *11*(3), 142.

Givens, M. e. (2013). *The black soldier fly how-to-guide*. Durham, NC: University of North Carolina.

Gorelick, S. (2016). How capitalism is changing Vermont's Maple syrup industry for the worse. *Shareable*. Retrieved from www.shareable.net/mapjam-3-0-join-us-to-put-sharing-on-the-map/

Goulson, D. (2019). The insect apocalypse, and why it matters. *Current Biology*, *29*(19), R967-R971.

Gowdy, J. a. (2013). The ultrasocial origin of the Anthropocene. *Ecological Economics*, 95, 137–147.

Green, A. (2008, November 27). Asian beetle spells death for maples so dear. *The New York Times*.

Haag-Wackernagel, D. a. (2004). Health hazards posed by feral pigeons. *Journal of Infection*, 307–313.

Han, F. A. (2012). From where did the Western honeybee (A pis mellifera) originate? *Ecology and Evolution*, *2*(8), 1949–1957.

Haraway, D. (2018). Staying with the trouble for multispecies environmental justice. *Dialogues in Human Geography*, 8(1), 102–105.

Heise, U. K. (2016). *Imagining extinction: The cultural meanings of endangered species*. Chicago, IL: University of Chicago Press.

Hinchliffe, S. a. (2006). Living cities: Towards a politics of conviviality. *Science as Culture*, *15*(2), 123–138.

Hinchliffe, S. a. (2008). People, animals and biosecurity in and through cities. In *Networked disease: Emerging infections in the global city* (pp. 214–227). Oxford, UK: Oxford.

Houston, D. e. (2018). Make kin, not cities! Multispecies entanglements and "becoming-world" in planning theory. *Planning Theory*, *17*(2), 190–212.

Human Rights Watch, a. U. (2003). *Small change: Bonded child labor in India's silk industry*. Bingley, UK: Emerald Group Publishing.

Jensen, N. H. (2019). "We will eat disgusting foods together": Evidence of the normative basis of Western entomophagy-disgust from an insect tasting. *Food Quality and Preference*, *72*, 109–115.

Jerolmack, C. (2008). How pigeons became rats: The cultural-spatial logic of problem animals. *Social Problems*, 72–94.

Jones, B. A. (2013). Zoonosis emergence linked to agricultural intensification and environmental change. *Proceedings of the National Academy of Sciences*, *110*(21), 8399–8404.

Jonsell, M. (2017). Black soldier fly, Hermetia illucens, invasive food? *Entomologisk Tidskrift*, *138*(3-4), 231–232.

Kimmerer, R. (2013). *Braiding sweetgrass: Indigenous wisdom, scientific knowledge and the teachings of plants*. Canada: Milkweed Editions.

Klein, J. (2007). *A vending machine for crows*. Dissertation, New York University, Interactive Telecommunications Program.

Lee, H. L. (1995). A case of human enteric myiasis due to larvae of *Hermetia illucens* (Family: *Stratiomyiadae*): First report in Malaysia. *The Malaysian Journal of Pathology*, *17*(2), 109–111.

Lenzen, M. e. (2020). Global socio-economic losses and environmental gains from the Coronavirus pandemic. *PloS One*, *15*(7).

Leong, M. R. (2018). Biodiversity and socioeconomics in the city: A review of the luxury effect. *Biology Letters*, *14*(5), n.a.

Lerner, H. a. (2017). A comparison of three holistic approaches to health: One health, EcoHealth, and planetary health. *Frontiers in Veterinary Science*, *163*(4).

Li, Q. e. (2011). From organic waste to biodiesel: Black soldier fly, *Hermetia illucens*, makes it feasible. *Fuel*, 90(4), 1545–1548.

MacInnis, G. a. (2019). Pollination by wild bees yields larger strawberries than pollination by honey bees. *Journal of Applied Ecology*, 56(4), 824–832.

Margulis, L. (2008). *Symbiotic planet: A new look at evolution*. Amherst: Basic Book, NYC.

McLain, R. J. (2014). Gathering "wild" food in the city: Rethinking the role of foraging in urban ecosystem planning and management. *Local Environment*, *19*(2), 220–240.

Mega, E. R. (2020). COVID has killed more than one million people: How many more will die? *Nature*. Retrieved September 30, 2020, from https://doi.org/10.1038/d41586-020-02762-y (Epub ahead of print, PMID: 32999443).

Mills, J. G. (2017). Urban habitat restoration provides a human health benefit through microbiome rewilding: The microbiome rewilding hypothesis. *Restoration Ecology*, *25*(6), 866–872.

Mooallem, J. (2006, October 5). Pigeon wars. *New York Times Magazine*.

Moore, L. J. (2013). *Buzz: Urban beekeeping and the power of the bee*. New York: New York University Press.

Murphy, B. A. (2009). How do we come to know? Exploring maple syrup production and climate change in near north Ontario. *Geography*, *2*.

Nading, A. M. (2013). Humans, animals, and health: From ecology to entanglement. *Environment and Society*, *4*(1), 60–78.

Nana, P. e. (2018). Black soldier flies (*Hermetia illucens* Linnaeus) as recyclers of organic waste and possible livestock feed. *International Journal of Biological and Chemical Sciences*, *12*(5), 2004–2015.

Nguyen, T. T. (2015). Ability of black soldier fly (Diptera: Stratiomyidae) larvae to recycle food waste. *Environmental Entomology*, *44*(2), 406–410.

Nigro, C. (2012, August 23). *How did the pigeon get to NYC?* Retrieved from New York Public Library, www.nypl.org/blog/2012/08/23/how-did-pigeon-get-nyc

Nimmo, R. (2015). Apiculture in the anthropocene: Between posthumanism and critical animal studies. In *Animals in the anthropocene: Critical perspectives on non-human futures* (pp. 177–200). Manchester: University of Manchester.

Norton, B. A. (2019). Urban meadows as an alternative to short mown grassland: Effects of composition and height on biodiversity. *Ecological Applications*, *29*(6).

O'Callaghan-Gordo, C. a. (2020). COVID-19: The disease of the Anthropocene. *Environmental Research*. Retrieved May 15, 2020, from https://dx.doi.org/ 10.1016/j.envres.2020.109683.

Oldroyd, B. P. (2012). Domestication of honey bees was associated with expansion of genetic diversity. *Molecular Ecology*, *21*(18), 4409–4411.

Osa, O. (2013). Pulke, maple syrup, and palm wine: International and multicultural literatures for cross-cultural learning and understanding. *Humanities Education*, *11*, 43.

Panelli, S., Epis, S., Cococcioni, L., Perini, M., Paroni, M., Bandi, C., Drago, L., & Vincenzo Zuccotti, G. (2020). Inflammatory bowel diseases, the hygiene hypothesis and the other side of the microbiota: Parasites and fungi. *Pharmacological Research*, *159*, 104962.

Poe, M. R. (2013). Urban forest justice and the rights to wild foods, medicines, and materials in the city. *Human Ecology*, *41*(3), 409–422.

Poe, M. R. (2014). Urban foraging and the relational ecologies of belonging. *Social & Cultural Geography*, *15*(8), 901–919.

Potts, S. G. (2018). Robotic bees for crop pollination: Why drones cannot replace biodiversity. *Science of the Total Environment*, *642*, 665–667.

Pozdnyakova, N. e. (2018). The degradative activity and adaptation potential of the litter-decomposing fungus Stropharia rugosoannulata. *World Journal of Microbiology and Biotechnology*, *34*(9), 133.

Quammen, D. (2012). *Spillover: Animal infections and the next human pandemic*. New York: W. W. Norton & Company.

Ryznar, R. M. (2001). Using remotely sensed imagery to detect urban change: Viewing Detroit from space. *Journal of the American Planning Association*, *67*(3), 327–336.

Sarpong, D. e. (2019). Biodegradation by composting of municipal organic solid waste into organic fertilizer using the black soldier fly (*Hermetia illucens*) (Diptera: Stratiomyidae) larvae. *International Journal of Recycling of Organic Waste in Agriculture*, 8(1), 45–54.

Scoones, I., Dry, S., Leach, M., Forster, P., Wilkinson, A., Bloom, G., . . . & Standing, H. (2018). *Free-book: Pathways to health and sustainability*. Oxfordshire, UK: Routledge.

Shackleton, C. M. (2017). Urban foraging: A ubiquitous human practice overlooked by urban planners, policy, and research. *Sustainability*, 9(10), 1884.

Simone-Finstrom, M. e. (2016). Migratory management and environmental conditions affect lifespan and oxidative stress in honey bees. *Scientific Reports*, 6.

Spranghers, T. e. (2017). Cold hardiness of the black soldier fly (Diptera: Stratiomyidae). *Journal of Economic Entomology*, *110*(4), 1501–1507.

Sturges, M. (2018). "Bleed on, blest tree!": Maple sugar georgics in the early American Republic. *Early American Studies: An Interdisciplinary Journal*, *16*(2), 353–380.

Swinscoe, I. e. (2019). Seaweed-fed black soldier fly (*Hermetia illucens*) larvae as feed for salmon aquaculture: assessing the risks of pathogen transfer. *Journal of Insects as Food and Feed*, 5(1), 15–27.

Tomberlin, J. K. (2020). Black soldier fly from pest to "crown jewel" of the insects as feed industry: An historical perspective. *Journal of Insects as Food and Feed*, 6(1), 1–4.

Valentín, L. e. (2013). Mycoremediation of wood and soil from an old sawmill area contaminated for decades. *Journal of Hazardous Materials*, *260*, 668–675.

Villanueva-G, R. D.-U. (2005). Extinction of *Melipona beecheii* and traditional beekeeping in the Yucatán peninsula. *Bee World*, 86(2), 35–41.

Villazana, J. a. (2019). Tolerance of immature black soldier flies (Diptera: Stratiomyidae) to cold temperatures above and below freezing point. *Journal of Economic Entomology*, *112*(6), 2632–2637.

Waltner-Toews, D. (2017). Zoonoses, one health and complexity: Wicked problems and constructive conflict. *Philosophical Transactions of the Royal Society B: Biological Sciences*, *372*.

Wang, Y.-S. a. (2017). Review of black soldier fly (*Hermetia illucens*) as animal feed and human food. *Foods*, 6(10), 91.

Wasko, N. J. (2020). Multiple sclerosis, the microbiome, TLR2, and the hygiene hypothesis. *Autoimmunity Reviews*, *19*(1).

Winfree, R. e. (2007). Native bees provide insurance against ongoing honey bee losses. *Ecology Letters*, *10*(11), 1105–1113.

Zhang, A. (2020). Circularity and enclosures: Metabolizing waste with the black soldier fly. *Cultural Anthropology*, *35*(1), 74–103.

Zimmer, C. (2020, April 24). Welcome to the virosphere. *The New York Times*.

Part III

Urban ecosystem justice pedagogy

8 The radix experiment

The Radix Ecological Sustainability Center[1] is a nonprofit urban environmental education center located in the South End of Albany, New York, cofounded by me and Stacy Pettigrew. Since its beginning in 2010, Radix has maintained a demonstration site of sustainable tools and technologies that are designed to teach local residents, with a special emphasis on youth, ways to have more local access and control over essential resources such as food, water, energy production, and waste management. By doing so, Radix hopes to empower citizens with the knowledge, tools, and technologies necessary for replicating these systems in their own homes and communities, with the larger goal of promoting a city overall that is more just, healthy, and resilient in both its social and ecological dimensions. Radix seeks, through its advocacy, pedagogy, and practice, to explicitly recenter the social pillar of sustainability, demonstrating how issues of equity, access, fairness, and justice pertain to matters such as food, climate, soils, water, air, waste, and non-human life. By doing so, we seek to promote an agenda of urban ecosystem justice that challenges both urban ecological alienation and the neoliberalization of sustainability. We are aligned philosophically and politically with the broader environmental justice movement and situate our work within a just sustainabilities framework (Agyeman, 2003), advocating for the elimination of environmental risks and the equitable distribution of ecological amenities. It is our belief that there should be neither ecological sacrifice zones nor sacrificial people, despite the historical patterns of systemic race and class-based oppression that have created them (Scott, 2017). At the same time, we work in coalition with numerous community-based organizations to address issues of racial, food, and climate justice within the South End neighborhood. We seek to take leadership from local communities of color who are most aware of the causes and consequences of the inequalities and oppressions we wish to overcome. This is facilitated through regular meetings with community coalitions and through involvement in the Innovation Blocks, a program that delegates resident "block captains" to engage in door-to-door communication and listening with residents about community concerns.

In this section, I will explain the numerous considerations that went into the design and crafting of the Radix Center's educational goals and how they relate to contributing to the ecological, social, economic, and political dimensions of the resilience of the surrounding urban environment. Furthermore, I will explore how

DOI: 10.4324/9780367858711-9

various design philosophies, such as ecological, regenerative, and permaculture design have influenced the creation of Radix and continue to inform its development. In this regard I hope to demonstrate how Radix, through its pedagogical practices, may be used as a staging ground for the overall concept of urban ecosystem justice.

Land acknowledgment

I recognize and honor Indigenous Nations as the traditional stewards of the lands and waters on which the Radix Center exists today. The Capital Region of New York State occupies lands built on the homelands of the Kanien'kehá:ka (Mohawk) Haudenosaunee and the Stockbridge-Munsee Band of Mohican Nation. I respect the sovereignty of these and all other indigenous nations surviving today and pledge to support the rights of these nations and the interests of indigenous peoples.

Background

The Radix Center is situated on a one-acre parcel in an inner-city neighborhood in Albany, New York. When it was first acquired, the site was a vacant parcel strewn with trash, abandoned cars, and covered in asphalt. Years of work spent removing debris and tearing up asphalt has gradually transformed the lot into a thriving sustainability demonstration site, complete with numerous garden beds and perennial agroforestry plantings. Systems contained at Radix include a solar greenhouse, aquaponics, microlivestock, rainwater collection, composting systems, mushroom cultivation, beehives, and alternative energy demonstrations. Strong emphasis is put on the interrelationships between these systems and their connections to each other, and how by-products from one system component become the nutrient or energetic input for another. For instance, food wastes that are brought on to the site are fed to chickens which in turn upcycle the wastes into eggs and manure. Eggs are distributed to local residents while wastes are composted and used as fertilizer in gardens. Garden vegetables are in turn distributed while garden weeds are returned to the chickens. In this way, we model how subcomponents function as a synergetic whole in an integrated agroecological farm system. In addition to the high level of interconnectivity on the site itself, a great deal of interchange is undergone with the local neighborhood and the urban system more broadly, in terms of both physical and social resources. We strive to make it known that we are not a museum, a historical reenactment site, or a "farm that just happens to be in the city". Our presence and work is situated and contextualized (Sze, 2018) within the context of Albany's South End and is directed toward addressing the needs and issues of the local community first and foremost.

We offer hands-on workshops and tours to area school groups and interested citizens throughout the Northeast. Since 2011, thousands of people have visited, including K-12 school groups from area schools, student groups and classes from universities throughout the Northeast, and many local and state government officials. While many of these visits are formally arranged, perhaps the most frequent and valuable among them are informal stop-ins by local neighbors interested in just visiting the chickens, get some salad greens, or relax by the pond. It is the repeat visits by children, in particular, that give local residents a sense of place, familiarity, and belonging at Radix.

The South End of Albany is a textbook case of a food desert neighborhood, a predominantly African American community where there is an abundance of liquor stores and corner stores selling highly processed foods but little in the way of fresh, affordable, and nutritious produce (McClintock, 2011). An examination of the forces of systemic racism that have created the disparity of food access in low-income communities of color has led to the creation of the term "food apartheid" (Brones, 2018), which aptly describes the conditions in the South End. Creating regenerative food justice and food sovereignty involves taking on systemic, often wicked issues that require multi-scalar collaborations. Radix partners closely with AVillage . . . Inc., a community advocacy organization, and Trinity Alliance, a community service provider to increase access to healthy food in Albany's South End. Our programming to increase access to nutritious, affordable food includes maintaining neighborhood gardens, holding low-cost farmers markets, and providing youth in our youth programs with examples of healthy dishes and snacks made from the food they have grown.

On the environmental justice front, AVillage and Radix are also currently conducting a comprehensive health survey at Ezra Prentice Homes with residents serving as outreach workers and SUNY School of Public Health volunteers conducting surveys. The survey results have led to advocacy that has resulted in a new, year-long half-a-million dollars air quality monitoring program for the South End conducted by the NYS Department of Environmental Conservation and studies by the Port of Albany and the Capital District Transportation Committee on the extent of diesel truck traffic on South Pearl Street and options for rerouting that traffic.

Radix has made a deliberate choice to locate in the downtown section of an urban area. Currently, more than 50 percent of the world's population is living in urban areas. As this percentage is projected to go above 60 percent by 2030, it shows that for the first time in history, we have become a predominantly urban species. This awareness makes it especially critical to focus on redesigning urban areas to make them more sustainable – essentially capable of meeting the needs of their residents while simultaneously increasing their ecosystem services. Part of this work will involve a reframing of cities as complex adaptive socio-ecological systems, the result of coevolutionary processes between the human and non-human aspects of the urban ecosystem (Alberti, 2003). Once cities have been identified as ecosystems, it becomes possible to analyze the components of them that cause them to be unhealthy or low-functioning.

Albany is a second-tier, post-industrial city with a relatively stable population and small urban-to-rural gradient and accordingly shares many common social, ecological, and economic characteristics with other shrinking cities (Herrmann, 2016). Successful systems developed at Radix, therefore, will be easily transferrable and replicable in other proverbial "rust belt" towns. Furthermore, with a small urban-to-rural gradient, there is good potential for facilitating urban–rural interchange with the hope of building solidarity between the respective populations.

Ecologically regenerative micro-industries

In acknowledgment of the fact that unemployment is widespread throughout inner-city environments, Radix also strives to develop models for what it refers to as "ecologically regenerative micro-industries" or "ERMIs". ERMIs would couple sustainable practices with an economic engine, seeking to provide jobs for people while promoting sustainable activities such as gardening, composting, fish farming, growing mushrooms, raising chickens, etc. Ideally, urban waste products such as food scraps, wood chips, cardboard, etc. could be intercepted form the waste stream and "upcycled" into marketable products. Furthermore, the organic by-products of these businesses could be composted and therefore used to promote the regeneration of degraded or contaminated urban soil systems. The overall concept for this economic model is further supported by the study of regenerative design (Cole, 2013).

The Radix Center manages a youth-centered farmshare operation, where high school aged youth are involved in the processes of both growing food and running a Community Supported Agriculture (CSA) program. Once a week, youth converge at Radix to partake in activities such as harvesting, bagging and weighing produce, as well as managing the accounts of CSA members with tablets. By combining agriculture and business learning skills, a diverse cross-section of youth are brought to the space where they have the opportunity to learn important life skills. The farmshare program is an example of an ecologically regenerative micro-industry that adds an important component of being socially regenerative as well.

One other micro-industry run from Radix is Community Composting Initiative (CCI). Once per week, paying customers in Albany place their food wastes in plastic buckets outside their homes for collection. Food scraps are brought to Radix where they are fed to chickens. By feeding scraps to chickens, the volume of material is rapidly reduced, while simultaneously reducing the amount of commercial feed that would otherwise need to be purchased and providing chickens with a diverse diet. Residual food scraps are then collected from the chicken yard and mixed with wood chips in compost piles where they are permitted to age and turn in to finished compost. Completed compost is then used to build soil in garden beds. In this regard, the composting operation diverts organic matter from landfills (which would otherwise contribute to methane gas production) and uses it to regenerate the health of damaged soils in the urban ecosystem.

The concept of ERMIs is fundamentally based on ancient practices. The discipline of historical ecology has shown us that certain ancient societies engaged in regenerative practices, like creating middens of organic matter that over centuries resulted in the creation of fertile soils and the enhancement of local biodiversity (Balée, 2002).

Political relationships

Being located in the New York State Capital was another deliberate choice on behalf of Radix. While Radix primarily believes in the effectiveness of grassroots, community-based activism, it also recognizes the limitations of this approach. By locating in a center of state government, Radix has the potential to be influential on a policymaking level. Radix is well situated to give tours and sustainability education programs to a number of lawmakers and policy workers. Groups that have visited Radix so far include members of NYSERDA, DASNY, EFC, DEC, the state senate, and Albany county legislature. Radix also works on a direct and local political level. As appointed member of the Sustainability Advisory Committee to Albany's Common Council, I work as a liaison between governmental and citizen organizations, hoping to dismantle the legal barriers that block the implementation of sustainable practices and to use government as a tool to promote the building of local socio-ecological resilience.

Radix and permaculture

More than any other design philosophy, Radix has been highly influenced by permaculture design. Permaculture is defined as "an ecological design science grounded in a fundamental recognition that economic viability and social justice are interrelated with functioning ecological systems. Permaculture guides the redesign of systems for production, consumption, and inhabitation according to this foundational viewpoint". Furthermore, it is a "global grassroots development philosophy and sustainability movement that encompasses a set of ethical principles and design guidelines and techniques for creating sustainable, permanent culture and agriculture" (Lockyer, 2013).

Permaculture is an agglomeration of three words: permanent, culture, and agriculture. Modeling its designs for agroecosystems, buildings, and communities on patterns observed in nature, permaculture views humans and human creations and activities as part of the natural world. Rather than focusing exclusively just on human creations such as agroecosystems, buildings and communities, permaculture emphasizes the interconnections among these creations, humans, and the natural world. "[Permaculture] is informed by a global view, maintains a strong tradition of technology and knowledge transfer across diverse areas and cultural traditions, and is fundamentally based on empirical observation and experimentation" (Lockyer, 2013).

Permaculture is further based on the assumptions that "the environmental crisis is real and of a magnitude that will certainly transform modern global industrial society beyond recognition. In the process, the well-being and even survival of the world's expanding population is directly threatened". Additionally, "the inevitable depletion of fossil fuels within a few generations will see a return to the general patterns observable in nature and pre-industrial societies dependent on renewable energy and resources" (Holmgren, 2002).

Based upon this critique, permaculture draws from both modern science and traditional ecological knowledge to design sustainable human settlements and production systems. Permaculture theory is based off three guiding ethical principles: Care for the earth, Care for people, and Fair Share. Further guidance is given by its twelve ecological design principles (Holmgren, 2002):

(1) Creatively use and respond to change (envisions possibilities and intervene in effective ways)
(2) Observe and interact
(3) Catch and store energy (harvest while it is abundant)
(4) Obtain a yield (make sure you are getting valuable results)
(5) Self-regulate; accept feedback (be open to modify dysfunctional behaviors)
(6) Use and value renewables (reduce dependency on scarce resources)
(7) Produce no waste
(8) Design from pattern to detail (observe natural–social patterns and apply them to design)
(9) Integrate (capitalize on how things work together)
(10) Use small, slow solutions (local resources & responses, manageable scale)
(11) Use and value diversity (diversity leads to greater resilience)
(12) Use edges; value the marginal (important things happen at the intersections)

The principle of "catch and store energy" is put to work in Radix's solar greenhouse. The greenhouse is designed to collect the heat of the sun during the prime solar heating hours of the day and to store it in sources of thermal mass (stone, earth, water) that are contained within the insulated boundaries of the greenhouse itself. During the night, when the air temperature inside the greenhouse begins to drop, the latent heat stored within this mass is released, thereby moderating the greenhouse's internal temperature. This has the effect of significantly reducing the amount of supplemental heating that is used to keep the greenhouse from freezing.

The principle of using small, slow solutions is reflected in Radix's emphasis of scalable systems. Radix acknowledges that it is completely unrealistic for the majority of urban residents to build a 20′ × 60′ greenhouse in their backyards. They instead emphasize that students at the center learn basic concepts and ideas that can be scaled down (in most instances) or potentially up accordingly to be able to fit whatever space they have access to. The majority of urban residents will not have access to any land space whatsoever. For them, it is important to have "parallel demonstrations" of systems that are appropriate for apartments, window boxes, or potentially rooftops. Such systems need to be small, lightweight, non-odiferous, and not require copious quantities of sunlight. These types of systems are also "nomadic" or capable of being detached and carried from place to place by transient renters. In addition, incrementalism in scaling systems up is always encouraged. Radix always recommends that people begin with small-scale systems which require minimal investment. Mistakes and failures incurred on small systems will result in greatly reduced consequences than on larger ones. Once a person gains familiarity, confidence, and fully understands the function and value of a system, it would then be appropriate to replicate it on a larger scale. Too often people will invest large amounts of resources into a large-scale project without having slowly developed the skills necessary to maintain it. This commonly results in accordingly large-scale disasters.

Radix attempts to integrate the principle of maximizing diversity when selecting both animal and plant varieties for cultivation. Genetically diverse populations tend to be more resistant and resilient to disease, as it is common that in diverse populations at least some of the members will be resistant to any particular disease. A massive shortcoming of industrial agriculture is its preference for cultivating monocultures or crops of genetically similar or identical plants and animals. When disease strikes a monoculture, it has the capability to cause significant harm. In this regard, diversity is defense against catastrophe. Additionally, by maximizing species diversity, a greater number of ecological niches are created that can in turn support a greater number of secondary or dependent species, such as pollinating insects.

In future, more research will be done to integrate permaculture principles into agroecological and sustainable agriculture practice. Furthermore, work remains to be done in diversifying the field of permaculture and making it more accessible to marginalized communities (Ferguson, 2015), as well as reconciling it with traditional ecological knowledge systems and potential anti-urban biases. Despite these issues, the field holds promise to be a key component, among many others, in building urban ecosystem justice.

Radix and resilience

One other paradigm that has influenced the goals of Radix is resilience theory. Resilience, as a scientific concept, has its first usage in the ecological sciences. Developed by C.S. Holling in the 1970s, ecological resilience challenged the previously held notion in empirical ecology (influenced by physics and mathematics) that ecosystems reached "climax" or equilibrium states. Resilience theory postulated that instead, ecosystems are constantly in a state of flux, moving from states of increasingly concentrated energy/resources to a state of disorganization, then a subsequent reorganization. This process is referred to as the "adaptive cycle" in resilience theory. In this regard, there is no singular "optimal state" for any ecosystem to exist in, but rather a cycle of phases in which the system can function and exist within in the face of disturbance and change (Walker, 2012).

System resilience is an important design criterion for Radix. To the fullest extent possible, systems at Radix have been designed to withstand disturbances and still retain function. Radix's aquaponic system is one example. Aquaponics is a system where fish and plants are grown together in a closed-loop recirculating loop where the wastes produced by fish become the nutrient inputs for plants. Essentially, it is trying to enhance the already inherent productivity of aquatic ecosystems by increasing the system's carrying capacity through artificial aeration and filtration. The drawback to aquaponics is that if a power failure were to occur, fish could start to die rapidly. An aquaponics system, therefore, can be designed along a continuum of resilience to efficiency, where fish contained in a system with a low stocking density could survive for longer in the case of a power failure. On account of the low stocking density, the fish in resilient aquaponic systems are also less stressed and susceptible to disease.

Adaptive capacity and transformation are two core principles that inform Radix's design and social mission. Radix sees itself functioning greatly as an educational hub, a distribution point of information, skills, tools, and genetic resources working toward the goal of creating a distributed, decentralized community-centered resilience throughout the city. By encouraging and supporting city residents to replicate sustainable systems in their own apartments, backyards, and neighborhoods, a redundancy of critical systems is created. For instance, if one city resident had all of their rabbits killed by a raccoon, it would be possible for them to restock their population by obtaining new rabbits from a neighbor. Additionally, rabbit raisers could occasionally swap male rabbits with one another and in doing so ensure that the city's domestic rabbit population as a whole does not suffer from genetic inbreeding. On yet another level, highly localized information and best practices can be shared amongst rabbit keepers throughout the city. Were any form of disaster to befall a city that resulted in a disruption of its food supply lines, the rabbits (along with the myriad of other crops and livestock raised in backyards/apartments) could serve as an emergency food supply for the city's populace. More in line with the vision of transformation and adaptive capacity outlined in resilience theory, small-scale decentralized livestock husbandry provides an alternative model to highly centralized, energy-intensive, and polluting food production models such as confined animal feedlot operations (CAFOs). In small-scale operations, animals can be fed human-generated waste products (food scraps), require little-to-no external energy inputs, and have their wastes recycled as compost which in turn assists in the regeneration of the health of damaged urban soils.

Broader connections

The Radix Ecological Sustainability Center teaches practical skills that can be applied to create environmental and economic sustainability. An emphasis is placed on issues of food security, health, and the remediation of contaminated soils. The Radix Center gives children and adults a hands-on educational experience to understand the main environmental issues we face today – and the tools to work toward positive solutions. We cover numerous topics related to waste reduction and reducing air, water, and soil pollution, including composting, renewable energy, rainwater harvesting, removal of impervious covers, and the impact of combined sewage overflows and eutrophication.

Through a partnership with the City of Albany Water Board, Radix has participated in the creation of a sign designed to teach urban residents about the relationship between impervious covers and urban watershed health. Depicting runoff reducing technologies at Radix, the sign demonstrates how ordinary citizens can participate in activities such as de-paving, rain garden construction, and rainwater harvesting, building beautiful landscapes while protecting the health of the Hudson.

Activities with second-grade classes at Giffen Elementary include a combination of activities taking place both at the school and the Radix Center (Figure 8.1). During cooler months, activities take place indoors in the classroom where students are engaged in growing microgreens, maintaining worm bins, observing aquaponics systems, and learning about alternative energy. During warmer months, students are able to take walking field trips to Radix to observe similar systems on a meso or neighborhood scale, developing a connection to a community farm and a "place-based" education.

Elementary age students learning about carnivorous pitcher plants and the role they can play in biological pest control. Appealing simultaneously to senses of disgust and fascination, the rows of dead insects inside the pitchers capture the children's attention and create a conversation starter about the differences between biological and chemical pest management. Students are challenged to think about how the use of toxic chemicals may in turn affect their own health when applied to food.

Figure 8.1 Ecological Literacy Learning

Source: Scott Kellogg

The Radix Center's Youth Employment Program is an afterschool and summer experiential learning youth employment program for high school students. The program meets after school once a week during the school year. During the summer the program is expanded as youth from the City of Albany's Summer Youth Employment program attend four days a week. We work closely with our community partner organization AVillage, Inc. during the expanded summer program. It is our goal that as youths attend the program they gain an in-depth interest and knowledge and will one day be project leaders as the program grows. College interns participate alongside the youth, creating opportunities for role model development and cultural exchanges.

The youth program teaches an appreciation of the urban ecosystem and the interconnections between its various environmental and social components. A central feature is assisting with the Radix Center's weekly Farm Share, a produce subscription service that provides a bag of fresh produce to 15–20 families. Firsthand experiences of soil preparation, planting, and harvesting connect youth to nutrient-dense food. The program gives youth job readiness skills and inspiration for green small business possibilities through direct participation. Youth also grow and harvest food for the AVillage Farmers Market, one of very few sources of fresh vegetables in the neighborhood. This provides opportunities to discuss the food system and access to healthy food.

We also spend significant time discussing numerous topics related to waste reduction and reducing air, water, and soil pollution, including composting, renewable energy, rainwater harvesting, removal of impervious covers, and the impact of combined sewage overflows and eutrophication.

During the summer program, one day a week the youth attend a workshop on urban biodiversity, watershed hydrology, waste, energy, the reduction of toxicity in soils and water, nutrition and healthy food preparation. The workshops include discussions and hand-on activities. A separate day of the week is reserved for field trip to a location that reinforces workshop topics. Past field trips included kayaking around Peebles Island, hiking, boating, and swimming at Grafton State Park, biking from the South End to Delmar on the new rail trail, launching a floating island wetland restorer into the Hudson River, and touring the County Waste recycling facility.

We also host numerous groups and hold monthly tours. Our goal is to conduct outreach to school to increase the number of groups visiting to learn about ecology and sustainability at the Radix Center.

It is our hope that as the youth build relationships of familiarity, respect, and concern with natural processes, they will share this knowledge with their peers and families, broadening the program's reach (Figure 8.2). Understanding the complex interactions between environment, society, and personal health challenges youth to examine the ways in which their community's well-being is impacted by environmental harms and empowers them to make positive changes in their own lives.

High school age youths are paid regular stipends either through grant funding or through the city's summer youth employment program to participate in the management of a community-supported agriculture program. Youth are responsible for gardening maintenance, planting, harvesting, and bookkeeping. By combining business skills and urban agriculture, youth can apply the skills and knowledge learned in the program to any number of ventures.

Figure 8.2 Participant in Radix Youth Program

Source: Scott Kellogg

Educational philosophy

Radix's educational philosophy is based on the idea that urban environmental education is an essential component of building sustainable socio-ecological systems. Traditional environmental education models focus primarily on wilderness preservation and the idea of nature as being something external to the city. Educators would focus on bringing students outside of urban environments with the attitude that nothing "natural" is within them that is worth studying. This deficit-driven model reinforces negative attitudes toward cities that can be internalized by youth (Thomashow, 2015).

In more recent years, coinciding with the emergence of the "ecology of the city" paradigm, there has been a greater focus on teaching about environmental and ecological processes underway in urban environments. Such pedagogical methods provide educational experiences for urban youth that are concrete, environmentally positive, and action-oriented. Furthermore, they take place in relevant contexts, have a long-term duration, provide support, follow-up, and reinforcement by role models. All these factors combined are more likely to bring about longer-lasting changes in environmental attitudes in youth (Matthews, 1995). Educational programs that fail to meet these criteria can have damaging consequences. As expressed by Agyeman,

> [E]nvironmental content and pedagogies which don't reflect multiple cultural perspectives, reinforce limited concepts of "the environment" and environmental protection, and circumscribe and legitimate a limited environmental discourse will not produce empowered and informed students, graduates and environmental leaders.
>
> (Grass, 2002)

Failure on the part of the environmental educator to be adequately teaching about the complexity of urban ecosystems may result in a privileging of attitudes and level of awareness drawn from the educator's own limited understanding. For this reason, it is highly important to be soliciting the experiences and knowledge of local residents and youth in the construction of a place-based urban curriculum. Doing so will give the educator a more complete understanding of the highly varied social, cultural, political, economic, and environmental aspects of their environment that they may otherwise be unaware of (Thomashow, 2015).

Recognizing the limited value of a single class visit, Radix is additionally engaged in attempting to integrate sustainability sciences into school curriculum and working to build small-scale sustainable systems, such as aquaponics and vermicomposting directly in the classroom. Understanding that teachers are under enormous pressure to teach in accordance with standardized curriculum, Radix attempts to seam our educational program as much as possible with the mandated curriculum. Radix emphasized to educators that "living laboratory" environments such as greenhouses and gardens are fantastic ways to teach STEM disciplines, as they synthesize physics, chemistry, biology, and ecology along with economics and social science (Krasny, 2009). While Radix is willing to work with the current enthusiasm for STEM-related topics, we are simultaneously critical of the reductionist and capitalistic forces that advocate for it, recognizing that the technological industry stands to profit off of a pretrained workforce. We advocate for a balanced approach to STEM education that equally values the insights of the social sciences and humanities, as well as community-based and traditional ecological knowledges (Wolfmeyer, 2013).

Students visiting Radix are encouraged to engage all of their senses while learning at Radix. While traditional curriculum places great emphasis on simply listening and watching, students at Radix are permitted to literally get their hands dirty by physically handling living systems such as water plants, snails, soil, worms, etc. When appropriate, they are also encouraged to taste different types of food as well. By appealing to different senses and learning styles, Radix hopes to reinforce learning at multiple levels, a pedagogical theory referred to as "multiple intelligence theory" (Gardner, 2011). Furthermore, students are encouraged to engage in "full cycle" learning (Sobel, 2004). In this regard, students are allowed to go through the full process of growing food from soil preparation, seed planting, plant cultivation, harvesting and food preparation, and completing the cycle by soil recycling/composting. In order to facilitate this, students are allowed to bring soil trays prepared and planted by them at Radix back to their classrooms for further cultivation.

By maintaining a solar greenhouse, Radix is able to affordably maintain an educational space that remains above freezing year round. This allows Radix to support plant growth and other biological processes throughout the year. From the perspective of students interested in studying biological processes, this is exceptional since the traditional school year is mostly the inverse of the growing season. Without a greenhouse, there would be reduced opportunities to otherwise study biological processes.

Ecojustice education

Radix's pedagogical approach is additionally informed by the idea of "Ecojustice Education", as presented by Martusewicz et al. (Martusewicz, 2014). Drawing heavily from the work of C.A. Bowers, the main tenants of ecojustice education can be summarized as follows:

- The recognition and analysis of cultural assumptions that undermine local and global ecosystems essential to life
- The recognition of entrenched patterns of domination that define people of color, women, the poor, and the natural world as inferior and less worthy of life
- An analysis of globalization/neoliberalism that has resulted in the exploitation of the southern hemisphere for the benefit of the North
- The recognition of environmental commons and relationships among people not requiring of money
- Earth democracies – rights of ecosystems to regenerate, well-being of future generations
- Commitment to pedagogy that teaches students to identify causes of social and ecological violence and to remediate them

The ecojustice education approach overall rejects what it sees as the false dichotomy between environmental and social justice issues, arguing that they must be viewed as having common cultural origins. It encourages the understanding that both ecological and social crises have intertwined cultural roots in the deep assumptions of modernity. Ecojustice views itself as a "pedagogy of responsibility", drawing from teachers, teacher educators, and community educators writing on the cultural roots of ecological crisis. It challenges students with such questions as "what are my just and ethical obligations to my communities?" and "what do we need to conserve, and what needs to be transformed?", while asking teachers to look for solutions to building just and ecologically sustainable communities among traditional indigenous and existing communities within industrial society. This teaching method is seen as running counter to mainstream education, with proponents claiming that "schools currently help to reproduce a culture of and economic system whose short-term profit motive and ideology of unlimited growth have created a society that dangerously overshoots the carrying capacity of the bio-systems depended upon for life".

Not only critical of mainstream education, ecojustice distinguishes itself from mainstream environmental education approaches, primarily from how it approaches human–non-human relationships. In a typical environmental education curriculum, "ecology is defined, for the most part, as the scientific study and management of natural systems assumed to be outside of human communities. Intersections among human social problems and ecological problems are generally ignored in the curriculum" (Martusewicz, 2014).

Furthermore,

> much of the dominant discourse in environmental education . . . has been based upon a technocentric approach to environmentalism which favors initiating young people into the concepts and skills needed for finding scientific and technological solutions to environmental problems without addressing their root social, political, and economic causes.
>
> (Fien, 1995)

This approach is further critiqued for being limited by only looking primarily at the behavior of individuals and not at societal systems and the cultures that support them. It is also believed that environmental education teaches catastrophic education, which results only in gloom, despair, and inaction. The insights and analysis of ecojustice education has been heavily influential to Radix and serves as a guiding template for our educational and activist goals.

Symbiotically with the ecojustice approach, Radix actively works to promote ecological literacy in the educational system. Ecological literacy is the ability to understand the organizing principles of natural ecological systems that make life on earth possible and to be able to apply those ideas for creating sustainable human societies (Orr, 1992). By framing education through the lens of ecological literacy, there can be attentiveness to complexity, nonlinearity, and unpredictability, thus challenging educators to consider the role environmental education has in fostering socio-ecological well-being. Additionally, ecological literacy programs incorporate responses to change, such as disaster and social transformations, thereby adding an important dimension to sustainable development. Furthermore, ecological literacy frameworks allow for ways to situate learning in natural resources management in cities and other anthropogenic environments. By incorporating notions of feedback loops and nested systems, ecological literacy frameworks enable questions of "how stewardship practices, environmental education, and ecological and social system health might reinforce each other, and thus suggests an 'ecology of environmental education'" (Krasny, 2009) consisting of citizen, nonprofit, and government partners. Ecological literacy movements focus on restoration of damaged urban ecosystems as a means of building social networks and community well-being. When urban residents are engaged in such activities as community gardening, community forestry, and watershed restoration, they are allowed to experience the health and restorative benefits of these activities. These practices allow people to see their impact on biological systems, the physical environment, and other ecosystem components (Krasny, 2009).

Radix believes that it is important for citizens, especially youths, to have a basic familiarity and relationship with the natural systems that make it possible for us to survive. Many urban youth, as a consequence of living in cities, tend to be disconnected and unfamiliar with the sources of their food, the origin of water, where energy comes from, how waste is managed, or what interspecies relationships resemble. Although small cities may be surrounded by farms and fields immediately outside their borders, if you do not possess a car these spaces may remain inaccessible to you. For this reason, it makes sense to integrate green, ecological, and agricultural spaces into high-density urban environments where residents may have ongoing relationships with the non-human systems there, fostering ecological literacy.

Ecological rift

In my experience as an urban environmental educator at Radix, I am often struck not only by the frequent lack of understanding and familiarity that many urban residents possess regarding ecological systems and functions. Consequently, urban residents commonly have little understanding or connection with the soils, waters, non-human life, and atmospheres in their cities. This separation is attributable to a constellation of factors in addition to fear that includes actual physical barriers such as walls and highways that block people's access to rivers and greenspaces. On a broader level, low levels of ecological literacy are due in part to the nature–society, urban–rural dualism that results for many urbanites in the perception of the environment as something outside of cities with which they have no connection. This phenomenon is well articulated by geographer Nathan McClintock through the concept of "metabolic rift" (McClintock, 2011), which further informs Radix's pedagogical approach. McClintock details the historical origins of rift scholarship and proposes that urban agriculture be used as a means for mending this division.

McClintock claims that according to Marx, the development of capitalism and the processes of urbanization have resulted in the alienation of people from the natural environment. These processes have resulted in the depletion of soils and an accumulation of pollution in places of industrial production, also contributing to promoting a false dichotomy between "city and country, urban and rural, humans and nature, obscuring and effacing the linkages between them". McClintock breaks the concept of metabolic rift into three primary subdivisions: ecological rift, social rift, and individual rift. In each instance, he gives examples of how urban agriculture could be used to mend broken nature–social connections.

Ecological rift, as stated, is the most commonly studied variation of the three. McClintock makes the case that capital's expansion creates a cycle of rifts where attempts to heal a rift in one location may result in a geographical displacement of ecological crisis to another locale. An example given by McClintock is the depletion of reserves of bird guano from South America driven by the depletion of soils in North America and Europe. This results in what amounts to a "shell game with the environmental problems [capitalism] generates, moving them around rather than addressing the root causes". The resource-intensive food production system described here becomes dependent on what are referred to as "subsidies" – "Sustaining social metabolism under a food production system that depletes rather than regenerates the resource base depends on both spatial and temporal 'subsidies' to the food web." The current system of industrial agriculture is dependent on these subsidies on a global scale in the form of petroleum and natural gas inputs. Similar to the centuries-old deposits of bird guano, these limited fossil fuel resources are based off subsidies that are millions of years old.

It is here that McClintock believes that urban agriculture has great potential to heal the metabolic rift by rescaling nutrient cycles and reducing food production's dependence on petroleum-based resources. One way to achieve this would be to recycle organic wastes back into soil, thereby creating a closed loop of nutrient cycling. For this reason, we make composting central to work and education at Radix.

McClintock then goes on to explore the concept of social metabolic rift. The defining element of social rift is the commodification of land and labor. The displacement of rural famers and herders through seizure of the commons led to a migration of people to urban areas to become industrial workers. This process continues to this day, primarily in the Global South, which McClintock states is the primary drive for urban agriculture. About 70–75 percent of people in Nairobi, he cites, engage in urban agriculture as a way to produce food. Small allotments in marginal spaces can help to supplement a family's income and diet. At the same time, however, large companies can justify paying their employees lower wages on account of the fact that urban agriculture acts as a social safety net.

In the Global North, urban agriculture has played a different role. While always present, McClintock argues, its relative prominence or obscurity has grown and waned correspondingly during times of austerity or prosperity. According to him, a "crisis discourse helped to obscure the subsistence role that UA has always played in urban landscapes, as well as to devalue UA in times of prosperity". During times of economic prosperity, urban gardens often become the victims of gentrification, bulldozed to accommodate new development.

McClintock goes on to explain how urban agriculture plays a role in mitigating social rift by de-commodifying land, labor, and food. As gardens function as sites of interaction between people of wide-ranging age and ethnic backgrounds, they strengthen community ties and help to partially return the means of food production to urban populations. On numerous occasions I have witnessed the power of gardens to create community bonds across a wide array of residents, brought together for the sharing of food and appreciation of the commons.

Lastly, McClintock addresses the question of what he labels as individual rift or alienation. This stems from people seeing themselves as being essentially external to the environment and being alienated from processes as basic as knowing where their food originates from. Individual rift is difficult to overcome on account of how deeply rooted the nature–culture division is, but nonetheless there exists good potential to overcome it through urban agriculture precisely because it arises on the level of individual consciousness. Urban agriculture can help to alleviate "nature deficit disorder" in children, defined as being "the shift from direct to 'increasingly abstract and symbolic' contact with the outside environment in the contemporary political economy limits affective cognitive and evaluative development in children, leading to a rise in childhood behavioural problems".

McClintock continues by asserting that by "physically working in the soil, sowing seeds, cultivating, harvesting and preparing foods" urban agriculture reengages individuals with their natural environment.

> Not only do experiences in the garden bring the urban farmer, gardener, or beekeeper into direct contact with the biophysical environment – soil, plants,

> water, sunshine, rain, worms, insects, birds – as prescribed by behavioural scientists cited above, but also allows him or her to experience and metabolize the surrounding landscape, transforming it into a product that he or she can consume.

This reconnection of labor to the biophysical environment can additionally help to overcome what McClintock calls the "balkanization of knowledge", or the division of the biophysical and social sciences. Radix has attempted to create a physical environment where urban residents may be in close contact with these natural elements, having the time and space to gain familiarity with them and to lessen individual, social, and ecological alienation.

Ecophobia

The lack of familiarity that many urban residents have regarding ecological processes often goes beyond merely ignorance, manifesting as "negative knowledge", or fear, surrounding the non-human world. This fear could also be partially explained by the way environmental education is taught. By teaching "catastrophism" in schools, it's possible to produce "ecophobia" – fear of ecological problems and the natural world (Sobel, 1996). Ecophobia, according to Sobel, is a result of the premature abstraction of the environment by asking children to deal with environmental problems on a scale and scope beyond their control. Studying issues such as rainforest destruction, climate change, acid rain, oil spills, toxic waste, etc. can have a negative effect on young children, whose sense of time, self, and place is still forming. Rather than being effective in motivating them to take action on these issues, it can result in children feeling only a sense of hopelessness and disempowerment. By seeming terribly widespread and beyond their control, young children may turn away from rather than step up to problem solving. Alternatively, Sobel advocates for instead focusing on cultivating a sense of love, respect, and connection between children and nature, particularly with local environments where children have the opportunity to study and interact with non-human life firsthand. Conventional environmental educational approaches, however, tend not to favor this practice. An example given is how teachers commonly prefer to teach curriculum about tropical rainforests indoors and through electronic media rather than to explore the forest or the meadow nearby. Once children near middle school and have developed proper cognitive skills that allow them to think in the abstract, Sobel believes it is appropriate to discuss environmental problems in greater detail. For younger children (below age eight), however, Sobel believes in teaching "love before knowledge".

While I fundamentally agree with Sobel's ideas about primarily trying to cultivate a sense of love and respect with the non-human for young children, there are other factors that should be considered when teaching in an urban environment. Most of Sobel's work has been conducted in rural and semi-rural environments, where bucolic landscapes are closer at hand. I believe it is appropriate to incorporate some discussion of environmental risk with young children living in cities, particularly in regard to issues of environmental contamination. Just as parents will teach their children to be careful when crossing the street and to be wary of strangers, it is important to inform children about the potential hazards of soil and air pollution. In my experience working with second grade children in Albany, most of them are already quite familiar with the realities of air pollution with either themselves or their family members suffering from asthma or other respiratory illness. To pretend as if these conditions do not exist and to focus only on an imagined pristine nature does children a disservice. More importantly, I believe, is for discussions about environmental problems to conclude with some proposed solutions-based action, even if the problem is far more complex for any single simple solution to resolve. By leaving children with an actionable solution, it can be possible to avoid hopelessness and disempowerment. Having conversations about urban atmospheres, in fact, can be part of a greater "place-based" learning strategy that looks at the urban environment in its totality (Sobel, 2004). With such an approach, it can be possible to talk about urban natures in both their positive and negative aspects.

Urban ecosystem justice at Radix

The Radix Center's environmental education program teaches ecological literacy to urban youth, a population that as a whole tends to be disconnected from the environmental processes that sustain society. Placed with a framework of "urban ecosystem justice", students are challenged to look at the city in its totality as an ecosystem, examining the interrelationship between social justice issues and the biophysical dimensions of the urban ecosystem: soil, water, airsheds, climate, waste cycles, and biodiversity. Questions of access and equity are applied to each of these areas within an environmental justice context, while at the same time being situated within a STEM-focused curriculum that is hands-on, engaging, and directly relevant to the lived experience of students. By centering the curriculum on urban ecosystems, students are shown how concepts such as nature and environment are of critical importance in the city. By being able to recognize and value ecological processes in their own communities, students gain a greater sense and appreciation for environmental protection as a whole and can draw connections and comparisons between ecological processes in both urban and rural locations.

Soil studies center on the topic of food, discussing a range of topics related to food access and food justice, exploring ways that nutrition impacts both the personal and environmental health of communities (Figure 8.3). Emphasis is put on local and organic food production, drawing connections between sustainable agriculture and reduced demands for chemical pesticides, fertilizers, and food transportation distances. Sections on integrated pest management, no-till farming, and composting further stress the role that sustainable soil management has on combatting soil erosion and fertilizer/pesticide runoff and its subsequent impact on water bodies. Attention is also given to the issue of soil contamination and the importance in educating youth about soil safety, testing, and remediation strategies.

Figure 8.3 Youth Learning About Worm Composting
Source: Scott Kellogg

By directly touching soil and playing with soil organisms, elementary students are permitted to indulge their interest in worms in a supportive environment. Many times (for good reasons related to soil contamination) urban youth are actively discouraged from touching soil. Direct engagement allows children to understand the difference between living soils and lifeless "dirt". Furthermore, children can overcome fear or reluctance to touch worms, organisms critical for soil health.

Water studies examine the urban watershed as a whole, asking how mutually beneficial relationships can be constructed between urban residents and the water bodies in their communities. Centering on the phenomenon of the combined sewage overflow, students are challenged to think of ways or actions on their part that can reduce the percentage of impervious urban covers and improve the health of water bodies. Activities include constructing rainwater collection barrels, building raised "rain garden" beds, kayaking trips in surrounding water bodies, and the construction of "floating islands" – constructed wetlands attached to buoyant media for use in remediating urban stormwater runoff (Figures 8.4 and 8.5). Lastly, aquaponics systems are used as a means to model aquatic ecosystems, serving as an "interim" means of providing nontoxic fish protein to urban communities while long-term work is done to restore the health of urban water bodies and fisheries.

Despite the fact that the existence of the Hudson was the only reason, ecologically and economically, for the creation of the city of Albany, most residents have little connection with the river today. This is largely due to the fact that the construction of Interstate 787 has cut off most people's access to the river, as well as the perception of the river as being contaminated. By engaging youth in the participatory construction of a "floating island" they are taught some degree of reciprocal respect, love, and concern with the well-being of the river.

Facilitating trips to the edges of the Hudson and arranging kayaking trips is a fun and participatory way to challenge the ecological alienation that prevent

Figure 8.4 Launching a Floating Island

Source: Scott Kellogg

Figure 8.5 Youth on a Kayaking Trip on the Hudson

Source: Scott Kellogg

youth from engaging directly with surrounding water bodies. Although many youth have never been in a boat before, most reported having a positive experience and enjoying the activity.

Studies of airsheds and climate draw the connections between local air pollution and urban climate justice, examining the ways that airborne industrial

Figure 8.6 Youth at DEC Air Monitoring Station
Source: Scott Kellogg

pollutants and vehicular emissions impact the health of urban communities both in the short term (asthma, emphysema) and in the long term (urban heat waves, rising sea levels). Students learn about the air quality index (AQI) and fly flags of corresponding colors as a means of raising community awareness of the issue. Furthermore, in Albany students will have the opportunity to work alongside DEC employees as they conduct ongoing monitoring of air quality with mobile sensing equipment. In the long term, students are encouraged to think of ways that cities can play a role in slowing climate change on a global level by reducing emissions of carbon dioxide and methane and by using urban plants and soils to sequester atmospheric carbon dioxide (Figure 8.6).

Being "intangible", air is a difficult medium to teach about and engage with despite it being essential for survival. Air monitoring stations, such as these, create a way to "visualize" air, even if only in numbers. The station is designed to provide real-time data and can engage youth in a combined ecological and technological process.

Studies on urban biodiversity stress the concept of "biocultural diversity", or interconnections between social justice and non-human life in the city. Using the garden and food production as a starting point, students learn about the role that insect pollinators play in food production and of the importance in creating habitat for them in underutilized spaces (reconciliation ecology). From there, students are introduced to the concept of novel ecosystems and develop an appreciation for synanthropic species such as pigeons and crows. By valuing these species and seeing them as something other than a nuisance, a sense of respect and concern can be cultivated for familiar species which then may be extended to others. Lastly, an appreciation for ruderal vegetation or "weeds" is developed, particularly as to the role they play in soil stabilization, carbon dioxide sequestration, water filtration, and habitat creation.

Part III conclusion

In this chapter I have discussed the creation and operation of the Radix Ecological Sustainability Center, using it as a real-world and grounded example of how an urban ecosystems justice curriculum may be actualized. The multitiered strategy of teaching youth at varying age levels and encouraging familiarity with the various socio-ecological and economic workings of a city system hopefully serves as a template or an inspiration for groups looking to replicate similar strategies in other locations. The concurrent operation of Radix has been an essential component of my research, functioning as an applied and practice-based counter to an otherwise theory-heavy framework. Importantly, creating Radix has, in effect, been "building the ship I wish to study"; in order to study urban ecosystem justice in an engaged, direct, and embodied way, it has been necessary to immerse myself in the process of creating it. Through the operation of Radix, I hope to address the twin issues of the exclusion of social sustainability and ecological alienation. By centering our work on environmental justice issues and attempting to look at how questions of equity and access apply to urban environments, we hope to make social sustainability explicit in our work.

Note

1 "Radix" is the Latin word for "root" and refers both literally to the roots of the plants being cultivated on site and metaphorically to the work of addressing the root causes of systemic issues.

Works cited

Agyeman, J. a. (2003). Toward just sustainability in urban communities: Building equity rights with sustainable solutions. *The ANNALS of the American Academy of Political and Social Science*, *590*(1), 35–53.

Alberti, M. e. (2003). Integrating humans into ecology: Opportunities and challenges for studying urban ecosystems. *BioScience*, *53*(12), 1169–1179.

Balée, W. a. (2002). *Advances in historical ecology*. New York: Columbia University Press.

Brones, A. (2018). Food apartheid: The root of the problem with America's groceries. *The Guardian*, *15*.

Cole, R. J. (2013). Regenerative design, socio-ecological systems and co-evolution. *Building Research & Information*, *41*(2), 237–247.

Ferguson, R. S. (2015). Grassroots engagement with transition to sustainability: Diversity and modes of participation in the international permaculture movement. *Ecology and Society*, *20*(4).

Fien, J. (1995). Teaching for a sustainable world: The environmental and development education project for teacher education. *Environmental Education Research*, *1*(1), 21–33.

Gardner, H. E. (2011). *Frames of mind: The theory of multiple intelligences*. London: Hachette UK.

Grass, R. a. (2002). *Second national people of color environmental leadership summit - Summit II*. Washington, DC: Summit II.

Herrmann, D. L. (2016). Sustainability for shrinking cities. *Sustainability*, 911.

Holmgren, D. (2002). *Principles & pathways beyond sustainability*. Hepburn: Holmgren Design Services.

Krasny, M. E. (2009). Applying a resilience systems framework to urban environmental education. *Environmental Education Research*, *15*(4), 465–482.

Lockyer, J. a. (2013). *Environmental anthropology engaging ecotopia: Bioregionalism, permaculture, and ecovillages* (Vol. 17). New York: Berghahn Books.

Martusewicz, R. A. (2014). *Ecojustice education: Toward diverse, democratic, and sustainable communities*. London and New York: Routledge.

Matthews, B. E. (1995). *Teaching and evaluating outdoor ethics education programs*. Retrieved April 17, 2017, from www.files.eric.ed.gov/fulltext/ED401097.pdf

McClintock, N. (2011). From industrial garden to food desert. In *Cultivating food justice: Race, class, and sustainability* (pp. 89–120). Cambridge: MIT Press.

Orr, D. W. (1992). *Ecological literacy: Education and the transition to a postmodern world*. Albany, NY: State University of New York Press.

Scott, D. a. (2017). "Sacrifice Zones" in the green energy economy: Toward an environmental justice framework. *McGill Law Journal/Revue de droit de McGill*, *62*(3), 861–898.

Sobel, D. (1996). *Beyond ecophobia*. Great Barrington, MA: Orion Society.

Sobel, D. P.-B. (2004). *Connecting classrooms & communities*. Great Barrington, MA: Orion Society.

Sze, J. e. (2018). *Sustainability: Approaches to environmental justice and social power*. New York: New York University Press.

Thomashow, C. e. (2015). Urban complexity and environmental education. In *Urban environmental education*. Ithaca, NY: Cornell Civic Ecology Lab. Retrieved from www.researchgate.net/profile/Angelique_Hjarding/publication/302877963_Urban_Planning_and_Environmental_Education/l

Walker, B. a. (2012). *Resilience practice: Building capacity to absorb disturbance and maintain function*. Washington, DC: Island Press.

Wolfmeyer, M. (2013). STEM, meet ecojustice. *Green Theory and Praxis Journal*, *7*(1), 32–46.

Conclusion to the book

From planetary ecological devastation, enormous global inequality, racial oppression, pandemics, climate change, growing political authoritarianism, and the replacement of social cohesion with technological addiction, the world may seem a dark place. In this book, I have tried to develop a novel integrated framework that could help to reconcile some of these social and ecological ills. Laying out a choppy mosaic (with crudely applied grout between the tiles), I have tried in Part I to present an assemblage of concepts that create the theoretical vertebrae of urban ecosystem justice. If they fail to come together as a cohesive whole, I could only hope that they may provide a roughly drawn roadmap for future scholars to fill in the blank spots on. Part II has delved into the nuts and bolts of urban ecosystem justice, dedicating a section to separate biophysical sphere – soil, water, air, waste (compost), and biodiversity. In each I have drawn linkages between the physical, chemical, biological, and ecological aspects of those elements with their social and cultural dimensions, illustrating the ways that matters of equity, fairness, race, and class pertain to human–non-human relationships and make urban ecology a field of interest and relevance. Finally, in Part III, I have attempted to show how urban ecosystem justice may be used as an educational framework by both comparing it to other environmental pedagogies and by showing how it has been operationalized at the Radix Ecological Sustainability Center. Throughout all the sections, I have tried to challenge the distinction between scholars, teachers, and activists, demonstrating a synthesized praxis of theory and application that is complex, informed, engaged, embodied, and grounded.

This book is far from all-encompassing – the topic of urban ecosystems and related social justice issues is huge, and much has been left out. Its shortcomings, however, are partially deliberate. I am eager to avoid creating a new universalizing theory, particularly when it pertains to the lived experiences of marginalized populations. Urban socio-ecological systems are dynamic patchworks, multiplicities, and pluralities that defy generalization – they are ongoing, transforming stories that are not one person's to tell.

What more, however, could have been included if word counts, time, and bandwidth had allowed? It would be interesting to explore urban ecosystem justice in more inwards and outwards dimensions. Looking inwards, I am intrigued by how social justice applies to the human microbiome (Ishaq, 2019), how the internal ecology of the human body is shaped by external socio-environmental factors, and how that in turn influences human health and behavior. Scaling up, can the concept of the holobiont or the hologenome be applied to a city, its human and non-human inhabitants, and their microbial symbionts? Can cities be retrofitted to increase urban microbial biodiversity and microbial rewilding equitably (Mills et al., 2020); (Mills, 2019); (Sariola, 2020)? Could cities, on a hologenomic level, be thought of as units for evolutionary selection? This would overlap somewhat with the field of ecologic niche and urban eco-evolutionary theory, which explores how constructed habitats influence the evolution of those living within them (Alberti, 2020), a field that is rich with quantitative data but has yet to engage significantly with justice-oriented social sciences.

Looking outwards, how can urban ecosystem justice be integrated and reconciled with Earth System Science (Steffen, 2020), contextualizing just sustainabilities within a holistic, planetary, and deep-time analysis in an appropriate "politics of scale" (Jon, 2020). Lest we not become tunnel visioned in our discussion of urban issues, it is critical to situate cities as entities within a larger, living, complex, and evolving planet upon which life is made possible by the interactions of its biotic and abiotic actors. Gaia theory, once regarded as a fringe, crackpot, new-age pseudoscience (Lovelock, 2000), is now enjoying a renaissance. Growth in complexity sciences and evolutionary ecology has provided conceptual tools to understand Gaia not as an individual sentient organism (as reductionist evolutionary scientists misunderstood it to be) but rather as the relative equilibrium maintaining emergent property of all life (particularly microbial) on earth interacting with its crust, oceans, and atmospheres (Crist, 2010); (zu Castell, 2019). Gaia is now benefiting from a new generation of systems thinkers that have taken up its whole-systems and animistic worldview (both literally and metaphorically) to comprehend and navigate the Anthropocene (Stengers, 2015); (Clarke, 2017); (Lenton, 2018); (Latour, 2019). Much work remains to be done in understanding Gaia and particularly how it relates to matters of urban ecological justice. The theory, as well as its associated concepts of tipping points, non-linearity, positive feedbacks, animism, and panpsychicism, has much to offer an urban ecosystem justice perspective.

Finally, another chapter could be written on the topic of urban–rural ecosystem justice. The fact that cities are created and supported by their hinterlands is well documented (Cronon, 2009) and much work needs to be done about what just and reciprocal relationships between urban and rural spaces would look like. The perception of cities as resource vacuums has been seized upon by regressive pundits to foster resentment toward city dwellers in rural populations, driving extreme political polarization in the United States. A progressive platform that builds interchange and solidarity between urban and rural populations will need to engage in the complex intersectionality of race and class, recognizing the reality of ongoing systemic racial oppression, the existence of deep economic inequalities, and the dynamic of ecological parasitism that can form between city and country driven by capitalist demand. I look forward to participating in this work.

Works cited

Alberti, M. e. (2020). The complexity of urban eco-evolutionary dynamics. *BioScience, 70*(9), 772–793.

Clarke, B. (2017). Rethinking Gaia: Stengers, Latour, Margulis. *Theory, Culture & Society, 34*(4), 3–26.

Crist, E. H. (2010). *Gaia in turmoil: Climate change, biodepletion, and earth ethics in an age of crisis*. Cambridge, MA: Massachusetts Institute of Technology Press.

Cronon, W. (2009). *Nature's metropolis: Chicago and the Great West*. New York: W. W. Norton & Company.

Ishaq, S. L. (2019). Framing the discussion of microorganisms as a facet of social equity in human health. *PLoS Biology, 17*(11).

Jon, I. (2020). Scales of political action in the Anthropocene: Gaia, networks, and cities as frontiers of doing earthly politics. *Global Society, 34*(2), 163–185.

Latour, B. a. (2019). Extending the domain of freedom, or why Gaia is so hard to understand. *Critical Inquiry, 45*(3), 659–680.

Lenton, T. M. (2018). Gaia 2.0. *Science*, 1066–1068.

Lovelock, J. a. (2000). *Gaia: A new look at life on earth*. Oxford: Oxford Paperbacks.

Mills, J. G. (2019). Relating urban biodiversity to human health with the "holobiont" concept. *Frontiers in Microbiology, 10*, 550.

Mills, J. G., Bissett, A., Gellie, N. J., Lowe, A. J., Selway, C. A., Thomas, T., . . . & Breed, M. F. (2020). Revegetation of urban green space rewilds soil microbiotas with implications for human health and urban design. *Restoration Ecology, 28*, S322–S334.

Sariola, S. a. (2020). Toward a symbiotic perspective on public health: Recognizing the ambivalence of microbes in the Anthropocene. *Microorganisms, 8*(5), 746.

Steffen, W. e. (2020). The emergence and evolution of earth system science. *Nature Reviews Earth & Environment, 1*(1), 54–63.

Stengers, I. (2015). Accepting the reality of Gaia. In *The anthropocene and the global environmental crisis* (pp. 134–144). New York: Routledge.

zu Castell, W. U. (2019). Gaia – A holobiont-like system emerging from interaction. In *Emergence and modularity in life sciences* (pp. 255–279). Cham: Springer.

Index

Note: Page numbers in italics indicate a figure and page numbers in bold indicate a table on the corresponding page. Page numbers followed by "n" indicate a note.

For Product Safety Concerns and Information please contact our EU representative GPSR@taylorandfrancis.com
Taylor & Francis Verlag GmbH, Kaufingerstraße 24, 80331 München, Germany

www.ingramcontent.com/pod-product-compliance
Lightning Source LLC
LaVergne TN
LVHW021123300926
849160LV00010B/836